# The Lighting Supervisor's Toolkit

*The Lighting Supervisor's Toolkit* guides readers through the Lighting Supervisor's production process with an emphasis on the importance of the collaborative nature of the role.

Lifting the veil on a process regularly learned on the job, this book offers a deeper understanding of the role of Lighting Supervisor and how to take lighting designs from dreams to reality. Readers will learn to communicate with designers, analyze drawings, plan installations, document decisions, supervise crews, and innovate out-of-the-box solutions.

Providing guidance for technically focused individuals seeking deeper understanding of the profession, *The Lighting Supervisor's Toolkit* is ideal for students and professional technicians looking to take on important leadership roles in theatrical and entertainment lighting.

**Jason E. Weber** is in his sixth season as Lighting Supervisor at Actors Theatre of Louisville. During his 13-year career, he has been the Master Electrician on over 90 professional productions. Prior to Actors Theatre, Jason served as Lighting and Sound Supervisor for seven seasons with the Merrimack Repertory Theatre and Master Electrician for four seasons with the Adirondack Theatre Festival. Jason has an M.A. in Theater Education from Emerson College and a B.F.A. in Theater from Marietta College.

# The Focal Press Toolkit Series

Regardless of your profession, whether you're a Stage Manager or Stagehand, The Focal Press Toolkit Series has you covered. With all the insider secrets, paperwork, and day-to-day details that you could ever need for your chosen profession or specialty, these books provide you with a one-stop-shop to ensure a smooth production process.

**The Assistant Lighting Designer's Toolkit**
*Anne E. McMills*

**The Technical Director's Toolkit**
Process, Forms, and Philosophies for Successful Technical Direction
*Zachary Stribling and Richard Girtain*

**The Production Manager's Toolkit**
Successful Production Management in Theatre and Performing Arts
*Cary Gillett and Jay Sheehan*

**The Director's Toolkit**
*Robin Schraft*

**The Properties Director's Toolkit**
Managing a Prop Shop for Theatre
*Sandra Strawn and Lisa Schlenker*

**The Costume Supervisor's Toolkit**
Supervising Theatre Costume Production from First Meeting to Final Performance
*Rebecca Pride*

**The Stage Manager's Toolkit, 3rd edition**
Templates and Communication Techniques to Guide Your Theatre Production from First Meeting to Final Performance
*Laurie Kincman*

**The Lighting Supervisor's Toolkit**
Collaboration, Interrogation, and Innovation toward Engineering Brilliant Lighting Designs
*Jason E. Weber*

For more information about this series, please visit:
https://www.routledge.com/The-Focal-Press-Toolkit-Series/book-series/TFPTS

# The Lighting Supervisor's Toolkit

## Collaboration, Interrogation, and Innovation toward Engineering Brilliant Lighting Designs

*Jason E. Weber*

Routledge
Taylor & Francis Group
NEW YORK AND LONDON

First published 2021
by Routledge
52 Vanderbilt Avenue, New York, NY 10017

and by Routledge
2 Park Square, Milton Park, Abingdon, Oxon, OX14 4RN

*Routledge is an imprint of the Taylor & Francis Group, an informa business*

© 2021 Taylor & Francis

The right of Jason E. Weber to be identified as author of this work has been asserted by him in accordance with sections 77 and 78 of the Copyright, Designs and Patents Act 1988.

All rights reserved. No part of this book may be reprinted or reproduced or utilized in any form or by any electronic, mechanical, or other means, now known or hereafter invented, including photocopying and recording, or in any information storage or retrieval system, without permission in writing from the publishers.

*Trademark notice*: Product or corporate names may be trademarks or registered trademarks, and are used only for identification and explanation without intent to infringe.

*Library of Congress Cataloging-in-Publication Data*
Names: Weber, Jason E., author.
Title: The lighting supervisor's toolkit : collaboration, interrogation, and innovation toward engineering brilliant lighting designs / Jason E. Weber.
Description: New York, NY : Routledge, 2021. | Series: The Focal Press toolkit series | Includes index.
Identifiers: LCCN 2020038129 (print) | LCCN 2020038130 (ebook) | ISBN 9780367504663 (hardback) | ISBN 9780367504656 (paperback) | ISBN 9781003049999 (ebook)
Subjects: LCSH: Stage lighting--Handbooks, manuals, etc.
Classification: LCC PN2091.E4 W39 2021 (print) | LCC PN2091.E4 (ebook) | DDC 792.02/5--dc23
LC record available at https://lccn.loc.gov/2020038129
LC ebook record available at https://lccn.loc.gov/2020038130

ISBN: 978-0-367-50466-3 (hbk)
ISBN: 978-0-367-50465-6 (pbk)
ISBN: 978-1-003-04999-9 (ebk)

Typeset in Times New Roman and Helvetica
by KnowledgeWorks Global Ltd.

*For Kim*

# Table of Contents

Acknowledgments — xi

## Part ONE
## Collaboration — 1

### Chapter 1: Meet and Greet — 3

Who Is the Lighting Supervisor? — 3
Working with Other Departments — 8
Connecting with the Lighting Designer — 12
The Lighting Team — 13

### Chapter 2: The Architect and the Engineer — 19

Lighting Designer As Architect — 19
Lighting Supervisor As Engineer — 21
Peer to Peer Collaboration — 24

### Chapter 3: The Lean, Mean, Lighting Team — 27

Leadership Is Empowerment — 27
Defining a Role — 30
Three Hat Philosophy — 32
Independent Together — 33
Grounding — 33

| | | | |
|---|---|---|---|
| **Chapter 4:** | Always Learning; Always Teaching | | 35 |
| | A Learning Expert | | 35 |
| | Leaders are Teachers | | 37 |
| | Learning Relationships | | 38 |

**Part TWO**
**Interrogation** — 41

| | | | |
|---|---|---|---|
| **Chapter 5:** | The Pre-Production Process | | 43 |
| | Onboarding | | 43 |
| | Design Meetings | | 47 |
| | The Plot Submission | | 49 |
| **Chapter 6:** | The Review and the Price Out | | 53 |
| | Plot Review | | 53 |
| | Plot Clean-Up | | 56 |
| | Price Out | | 61 |
| **Chapter 7:** | Electrical Planning | | 67 |
| | Circuit and Dimming Infrastructure | | 67 |
| | Dimmering the Plot | | 80 |
| | Hot Power and Data Infrastructure | | 86 |
| | Hot Power and Data Paperwork | | 94 |
| **Chapter 8:** | Documentation and Shop Prep | | 99 |
| | Load-In Documentation | | 99 |
| | Color and Template Prep | | 104 |
| | Rigging Planning and Paperwork | | 110 |
| | Shop Prep | | 113 |

| **Chapter 9:** | The Load-In | **117** |
|---|---|---|
| | Dancing Not Fighting | 117 |
| | Installation Best Practice | 118 |
| | Hang | 119 |
| | Circuiting | 121 |
| | Other Waves | 126 |
| | Troubleshooting | 127 |
| **Chapter 10:** | Focus | **131** |
| | Preparing for the Big Game | 131 |
| | Calling the Focus | 132 |
| | Technician Focus Standards | 137 |
| **Chapter 11:** | Tech, Performance, and Strike | **141** |
| | Take a Step Back | 141 |
| | Notes Calls | 142 |
| | Empowering the Crew | 146 |
| | Strike | 148 |

**Part THREE**
**Innovation** **151**

| **Chapter 12:** | Boy, Wouldn't It Be Cool if...? | **153** |
|---|---|---|
| | Pushing the Boundaries | 153 |
| | Understanding the Idea | 157 |
| | Planning and Budgeting the Solution | 158 |
| | Prototyping and Solution Approval | 160 |
| | That's Not the Intended Purpose | 161 |

**Chapter 13:** Asset Management and Season Planning        165

    Loving Your Gear        165
    Asset Tracking        167
    Capital Expenses        170
    Season Budgeting        171

**Chapter 14:** Your Turn        175

    Rules of Thumb        175
    Further Reading        176

Appendix – Example Production Paperwork        177

*Index*        183

# Acknowldgments

This book would not be possible without the unknown contributions of all the technicians, apprentices, and interns I have worked with in my career. You have been my teachers and my lab rats. The work we have done together has made me into the Lighting Supervisor that I am today. I hope that this book will help carry your legacy for years to come.

I particularly want to thank my current and former staff from the Lighting Department at the Actors Theatre of Louisville—Dani, John, Wylder, Tyler, Andy, Oliver, Jacqueline, Katie, Jon, Vicki, Siena, Seth, Jessie, Natasha, and Cheyenne. This book began as a standards and procedures guide for our department and ballooned into something way bigger. Thank you all for helping to wade through some of my crazier ideas to get to the good ones.

# Part One
## Collaboration

Throughout my career people have regularly asked me, "So, what is it that you do, again?" At first, it was mostly from family who didn't know what to make of my career choices. This was largely understandable, I'm sure most folks that aren't lawyers or doctors face that line of questioning. Yet, I started to notice that I got the same questions from folks that were supposed to be in the know—theatre patrons, donors, board members, lighting students, and even newly hired technicians. I even began to ask the question myself. What do I do? Why does it seem so nebulous? What makes my role essential to the process? On the surface, this job is very technical—plugging lights in and making sure they don't fall on people's heads or catch on fire. You can't ignore all that, it is a big part of the job. Still, that's not the thing that makes a Lighting Supervisor an essential member of the production team. They are a communicator, a translator, an empathizer, a leader, and a teacher rolled into one. So, what do I do? Let's get into it.

CHAPTER 1

# Meet and Greet

## WHO IS THE LIGHTING SUPERVISOR?

Lighting is a very intangible element of a theatrical production. We know it is there and we know it impacts the performance, but, to most, how and why are great mysteries. Still, the biggest mystery of all might simply be: "who?" Most people are surprised by exactly how many people are involved with the lighting process.

The Lighting Team is comprised of many different roles—each with distinct responsibilities—that must work in collaboration with each other and the rest of the production staff to make the work we see each night the curtain rises.

To start, most people familiar with theatre production are familiar with the "Lighting Designer" or "L.D." The Lighting Designer is the marquee name for the lighting of a show. They will be front and center on the title page of the playbill. It is their job to work with the director to interpret the play in terms of light. In many ways, the Lighting Designer is the "architect" of a production's lighting—the one who dreams up the final product. They will determine where the lights will go, what kind of lights they will be, what colors or textures will be needed, and ultimately when and how bright the lights will be turned on.

## LIGHT PLOT

The Light Plot is the primary visual means of conveying the lighting design prior to technical rehearsals. The Light Plot shows the location, type, color, and purpose of each fixture requested by the Lighting Designer. The Lighting Designer uses this documentation to begin the conversation with the Lighting Supervisor and other members of the production team. As the process goes on, the Lighting Supervisor will add to this drawing to convey technical information such as circuiting or data infrastructure detail.

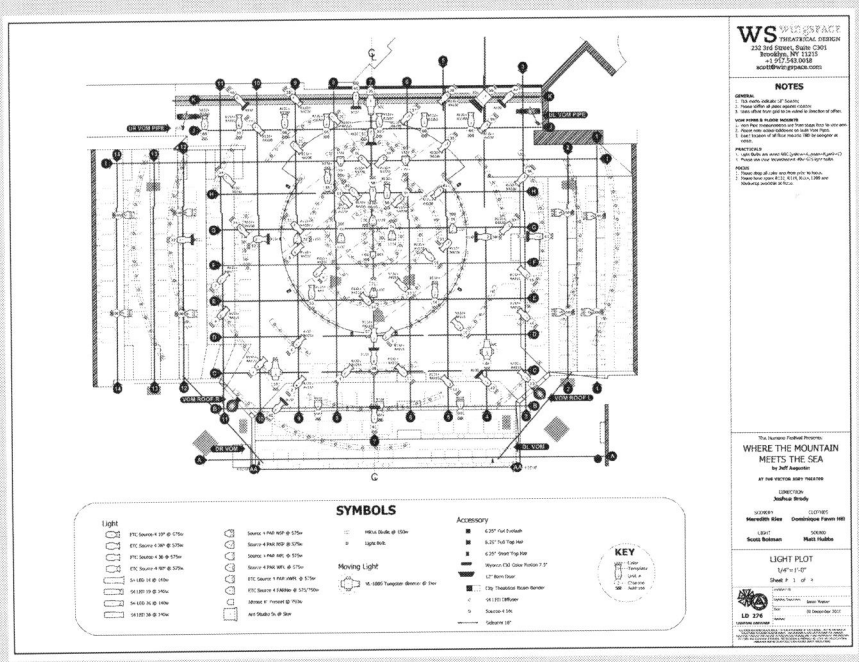

**Figure 1.1** Light plot for *Where the Mountain Meets the Sea* by Jeff Augustin, 2020 Humana Festival for New American Plays, Actors Theatre of Louisville; lighting by Scott Bolman.

Courtesy of Scott Bolman.

However, just like in building construction, there are many details about the functionality of a design that are needed to take that final product from paper to reality. Much of that work extends outside the role of the Lighting Designer. In fact, most light plots—a drafting plan that details the layout of the equipment in the venue—include a disclaimer that will read:

"The Lighting Designer is unqualified to determine the structural or electrical appropriateness of this design, and will not assume responsibility for improper engineering, construction, handling, or use. All materials and construction must comply with the most stringent applicable Federal and Local Fire, Safety, Energy, and Environmental Codes." (Shelley 2009)

If the Lighting Designer is not qualified to make these determinations, who is? Each production is going to need someone to "engineer" the lighting design. They will need someone to determine the structural and electrical soundness of the designer's light plot as well as lead the crew in the safe and proper execution of the design. This is the role of the Lighting Supervisor. They must partner with the Lighting Designer to make their artistic vision a practical reality.

Not every production will have someone with the title "Lighting Supervisor," but this does not mean that they do not have someone doing this engineering work. Many

## MASTER ELECTRICIANS AND THE ELECTRICAL TRADES

The term "Master Electrician" or "M.E." comes from the electrical trades. As part of the path for licensure, electrical tradespersons start as Apprentices, advance to Journeymen, then ultimately become Master Electricians. While this is the origin of the term, it is important to note that Master Electricians in a theatrical sense are not equivalent to Master Electricians in an electrical trades sense. In the electrical trades, Master Electricians are trained, licensed, and bonded which makes them "Qualified Personnel" according to the *National Electrical Code®*.[1]

Conversely, Master Electricians in the theatrical sense are not required to have any specific training or licensure to perform their roles. Consequently, there is a push among lighting professionals to move away from this terminology. On many Broadway shows, for example, the term Master Electrician is replaced with "Production Electrician." Additionally, organizations like the Entertainment Services and Technology Association (E.S.T.A.) have started a certification program for "Entertainment Electricians" as part of their Entertainment Technician Certification Program (E.T.C.P.).

While these other terms are growing in use, "Master Electrician" and "Electrician" are still very common in today's industry. For simplicity, when I use the term "Master Electrician" and "Electrician" here it is exclusively in the theatrical sense with a full acknowledgement of this problematic background.

[1] NFPA 70®, *National Electrical Code*®, and NEC® are registered trademarks of the National Fire Protection Association, Quincy, MA.
Please note that this permission is restricted to the use of the language referenced and does not include any permission to use the NEC Logo or any other NFPA logo or indicia, which may be used solely with the express written permission of NFPA.

companies use the terms "Master Electrician," "Head Electrician," or "Lighting Director" instead. These terms have similarities and are often used interchangeably by producers. However, among lighting professionals there are clear distinctions on the expected responsibilities that go with these titles. In understanding exactly who a Lighting Supervisor is and how they fit into the production process, it is important to understand more about these other common titles.

The "Master Electrician" is the most common. The Master Electrician—or sometimes "Production Electrician"—is the person in charge of engineering the lighting design for a specific production and generally does not have any obligation to the producers or the venue outside of the safe and functional installation of a specific production. In many cases, these individuals are freelance technicians who move from one production to the next in the area that they live. In a basic sense, a Lighting Supervisor is also a Master Electrician. However, the Lighting Supervisor has additional responsibilities that extend

## PRODUCERS VS. VENUE

Often, producers get conflated with their venues. If someone, for example, went to see a play at the Merrimack Repertory Theatre in Lowell, Massachusetts, they would probably say, "I saw a play at Merrimack Rep last night!" and not acknowledge that the performance was held in a venue called "Liberty Hall."

However, it is important to know the difference between a producer and a venue to understand how theatre is made. For example, if the same patron above went to see a show at Theatre Row in New York City, loved it, but didn't know that it was produced by Mint Theatre Company—which focuses on lost or forgotten plays—they could potentially be surprised if they went back to the venue and saw a show produced by Ma-Yi Theatre Company—which focuses on new works by Asian-American writers—thinking that a single company produced out of that venue.

To complicate matters further, many large theatre companies operate multiple venues. For example, Actors Theatre of Louisville in Louisville, Kentucky, has three venues—The Pamela Brown Auditorium, The Bingham Theatre, and The Victor Jory Theatre. If a person saw a play in the Bingham Theatre and hated it because they did not like theatre in the round, they may avoid all productions by the producer because they did not realize that the producer also operates venues with different seating configurations.

As a lighting professional, it is important to understand this relationship. If the producer does not operate a venue, the producers' lighting staff will have to interface with the venue's staff. Similarly, if a company operates a venue that "presents" the work of other companies, the venue's lighting staff may need to coordinate with the producer's lighting staff instead of directly with a designer.

beyond the scope of a single production and into the larger structure of the producing organization or venue. Despite this distinction, it is common for the terms to be used interchangeably. Many regional theatre companies with only one venue, for example, elect to call their lighting department manager, "Master Electrician," while many companies with multiple venues have "Master Electricians" for each venue that report to a central "Lighting Supervisor" who serves as the department manager.

Another common role is "Head Electrician." This is most often a term used for the person in charge of the lighting for a particular venue and is often associated with someone who is a member of the International Alliance of Theatrical Stage Employees (I.A.T.S.E.), a labor union that covers many stage technicians and designers. A Head Electrician's primary focus is the functionality of the venue and the equipment it owns. In some cases, a production will have both a Head Electrician and a Master Electrician. If operating within this scenario, the Master Electrician works with the Lighting Designer to focus on the engineering of a production, whereas the Head Electrician works with other lighting technicians to install the production into the venue. In cases where the producing company does not employ a lighting staff member, the venue's Head Electrician will also perform the work of the Master Electrician and function as a Lighting Supervisor because they are managing the venue as well as the lighting needs for a specific production. The primary difference between a Lighting Supervisor and a Head Electrician is their union affiliation. A Lighting Supervisor is nearly always a non-union manager, whereas a Head Electrician is almost always a member of the union who leads and works alongside other union technicians.

The final common role is "Lighting Director." Lighting Directors occupy a unique role because they have many of the same responsibilities as a Lighting Supervisor, but usually work more directly with the Lighting Designer on the artistic work of the production and supervise other staff members who work more directly on the technical work. Sometimes Lighting Directors will function as resident Lighting Designers who also manage the long-term goals of the organization and have staff Master Electricians who perform much of the engineering of specific productions. Lighting Directors are less common in theatrical settings but dominate dance and opera markets. Additionally, it is common to find Lighting Directors on touring shows. In that context, they adapt the touring design to meet the limitations of the individual venue stops and work with the venues' chief lighting technicians (often a union Head Electricians) to coordinate the installation. The big difference between a Lighting Supervisor and Lighting Director is the relationship with the designer. The Lighting Supervisor is exclusively a designer's technical collaborator, while a Lighting Director also provides artistic assistance. For example, a Lighting Director may function as an Assistant Lighting Designer during the technical rehearsal process or run focus calls for rotating repertory companies.

In film, the roles and relationships are quite a bit different. There is generally not a Lighting Designer on a film, but rather a "Director of Photography" or D.P. The D.P. will determine the lighting as well as other aspects of the filming process such as camera operation and shot selection. The D.P. works with the Gaffer who is essentially the Master Electrician on the film set. As films do not generally have the same recurring needs as theatrical companies and venues, the operational duties of a Lighting Supervisor are not needed on a film set. While some of the information in this book is useful for gaffers, the demands of the film industry are extremely different than those of live theatrical events, so readers in those industries should also explore many of the great books on lighting for film like *Set Lighting Technician's Handbook* by Harry C. Box.

## RESIDENT THEATRE COMPANIES

Prior to the mid-twentieth century, theatre in the United States was comprised primarily of touring shows and Broadway shows. Audiences in cities outside of New York would wait until a show traveled to them. During the 1950s and 1960s, there was a movement to create theatre companies in cities around the country. Each theatre was to have a resident company of artists that made theatre specifically for the community they were situated in. While the movement has evolved in many directions, the term "Resident Theatre" is still used to refer to theatre companies that produce a season of work geared toward the community they reside in with some semblance of a resident company of artists, technicians, and managers. The most well-known resident theatres are members of The League of Resident Theatres (L.O.R.T.).

The biggest distinction between producers who are resident theatres and other producers is in the mission. A resident theatre—usually also a nonprofit organization—works to serves their audience and community. The productions themselves are the conduit through which they serve. Other producers—many of whom are commercial—are focused on the productions in and of themselves. This can be for profit reasons, as is the case with Broadway producers, or for artistic reasons as is the case with some ensemble groups that make theatre without a specific audience in mind.

Understanding the type of producer you are working for is important. A Lighting Supervisor in a resident theatre needs to be able to align the goals of each production with the larger needs and mission of the organization, whereas production-focused Lighting Supervisors do not have those obligations. In fact, these producers often forego hiring Lighting Supervisors specifically and prefer production-specific Master Electricians or Head Electricians.

## WORKING WITH OTHER DEPARTMENTS

One of the key aspects of the Lighting Supervisor role is their relationship with the company they work for. While their specific production work is crucial, the work they do in service of the larger organization is equally important. Understanding the hierarchy and structure of theatre companies and of production teams is useful in understanding how the Lighting Supervisor interacts with their colleagues within an organization.

Most resident theatre companies in the United States follow the same basic structure. A Board of Directors appoints one or more executive leaders. In some organizations, the executive leadership is divided between "administration" and "artistic." In other organizations, that leadership is unified under one person. If divided, the production departments nest under the "artistic" side. This is usually led by the "Artistic Director." The Artistic Director oversees the selection of plays in a company's season as well as the hiring and supervision of the artists that make the plays including playwrights, directors, designers, and actors.

Directly reporting to the Artistic Director is the "Director of Production" or "Production Manager." The Director of Production manages the technical needs of all productions as well as maintains the company's production capacity for the future. Those technical needs are divided into the various production departments. Each production department is headed by a technician-manager—one of which is the Lighting Supervisor—who brings expertise in their area as well as provides logistical coordination of their departmental assets and staff. The individual members of the production departments report to their respective department head.

The relationship the Lighting Supervisor has with the Director of Production is particularly important. Like most relationships, a successful one here needs to be built on trust and mutual respect. Historically, Directors of Production have been promoted from their role as one of the production department heads. In this tradition, some companies may have a Director of Production that used to be a Technical Director, Production Stage Manager, or even a Lighting Supervisor. This results in Directors of Production with a very deep understanding of only some aspects of theatre but with a significant amount of time spent in the industry and experience to lean on. Today, many Directors of Production are training specifically for that role and have a broader understanding of all areas of theatre, while sacrificing the experience and depth of knowledge gained from years spent in a particular role.

Regardless of the background of any particular Director of Production, more often than not, they will not have the same depth of expertise in lighting that the Lighting Supervisor has. Therefore, the Lighting Supervisor needs to position themselves as the Director of Production's lighting consultant. This requires the Lighting Supervisor to develop clear recommendations and deliver them without expectation. In turn, the Director of Production needs to be able to bring the humility of knowing that they need to trust the validity of the recommendation. For example, the company may own a series of lighting fixtures that were just discontinued. It is unlikely that a Director of Production would be aware of any discontinuation notice or what the impact of it would be on the company. The Lighting Supervisor would be best poised to understand the context and provide different options on how the company could respond. The Director of Production

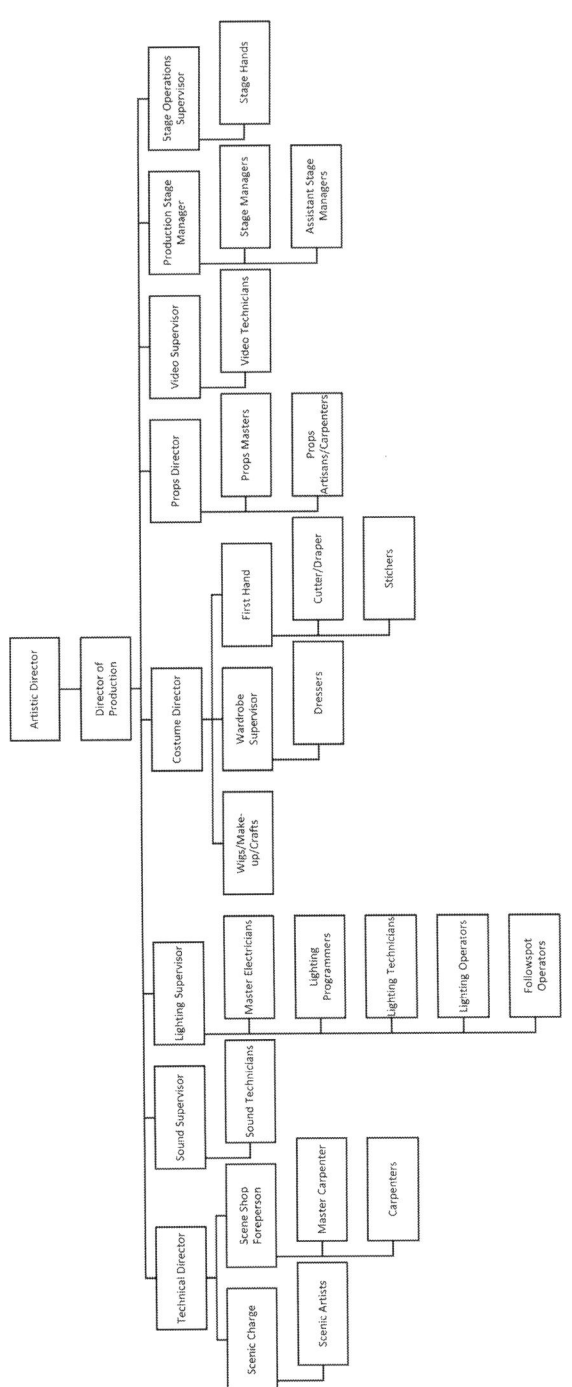

**Figure 1.2** Typical structure of a regional theatre in the United States.

would be able to use that information in connection to the overall needs of the organization to develop a response plan and make the final decision on how to proceed.

It is equally important for a Lighting Supervisor to maintain good communication with the managers of other department areas. Rarely will the lighting for a production be able to be installed without any coordination from other departments—the Sound Department may wish to install speakers in the way of lights; the Scenic Department may take out the floor on the day moving lights are intended to be roped up to the catwalks; the Costume Department may need to build a light-up hat the day before the focus call. A Lighting Supervisor must keep in communication with other departments from the very beginning of the process through the end.

This is especially true with cross-departmental collaborations. There will likely be instances on each production where an element of the show will need to be created by multiple departments. Frequently, for example, the Lighting Department collaborates with

**Figure 1.3** A "Practical"—a plastic birthday cake with remote controllable electric candles. The cake was built by the Props Department and the wiring was done by the Lighting Department.

the Props Department on "practicals." These practicals—props that light up—require the Props Department's artistic attention to detail to ensure the item looks exactly how the Scenic Designer intends, but also the Lighting Department's electrical know-how to ensure that the light inside works correctly, reliably, and to the Lighting Designer's specification. It is easy for one department or the other to decide to take on the entirety of such a project, but in doing so they deny the other department's expertise and risk a sub-par final product. If both departments honor the other's expertise, the final product will inevitably be more impactful.

## CONNECTING WITH THE LIGHTING DESIGNER

Honoring expertise is the backbone of any professional relationship and that is especially evident when discussing the relationship between the Lighting Supervisor and the Lighting Designer—the most significant relationship the Lighting Supervisor has. It is very important to understand that the Lighting Supervisor and the Lighting Designer are collaborative peers. Just like the Lighting Supervisor and the Props Director in the above practical example, each brings a specific expertise and role to the process.

When considering the Designer-Supervisor relationship, it is important to note that the Designer works outside of the structure and hierarchy of the theatre company. Does being outside of the hierarchy make them above it? Not at all. They are instead part of a production hierarchy that nests into the theatre company's structure for each production they do. For each production, designers report to a production's director on matters of artistic intent and to the Director of Production on technical matters. In this way, the Designer and the Lighting Supervisor share a supervisor. If we were to draw these nested hierarchies out—as in Figure 1.4—we can see that the Lighting Designer and Lighting Supervisor function as hierarchical peers.

This peership is important because each have different core concerns. The Lighting Designer needs to focus on the artistic success of a production, whereas the Lighting Supervisor needs to focus on the long-term success of a company's mission. This long-term success includes ensuring that the assets of the company are used correctly. For example, a Lighting Designer may want to turn off the fans in all the scroller units in a production to give the play an absolute quiet moment. They argue that the maximum artistic success of that moment would only be achieved if that silence existed. They would not have to consider how that choice would impact the life of the scrollers and their use in future productions because they are only working on this one production. The Lighting Supervisor would have to weigh this artistic need with the long-term health of the equipment.

It's easy to consider this lack of concern as callous, but in order for the artistic elements of the Lighting Design to be successful, the Lighting Designer needs to be able to consider each in an absolute and ideal sense. When a Lighting Designer devotes time to the techni-

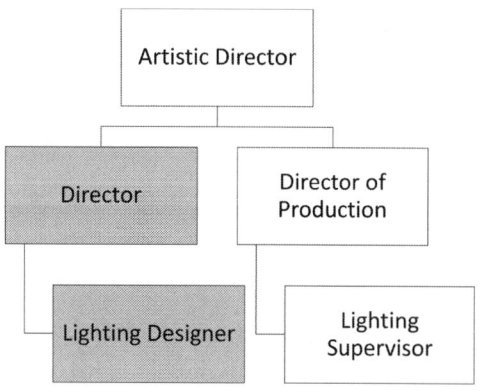

**Figure 1.4** The Lighting Supervisor and the Lighting Designer as hierarchical peers.

cal details or long-term outcomes of the theatre company, they are unable to focus on creating a brilliant design. The Lighting Supervisor needs to be able to come in and handle the practical considerations of the design so that the designer is free to focus on their role. When the art is at odds with the practical, the Lighting Supervisor collaborates with the Lighting Designer to develop solutions. In this example, the Lighting Supervisor would have to balance the Designer's desire for silence with their desire for using the scrollers in that scene. Through discussion, they could determine if it would be equally successful to add a new system of lights to achieve the color goals for the scene without the fan noise of if there was another solution.

## THE LIGHTING TEAM

With lighting divided into the artistic and the technical, so too is the Lighting Team. Often more people are needed than just a Lighting Designer and a Lighting Supervisor. Each person leads teams that work in tandem to balance each other and ensure a dynamic production that both looks amazing and is structurally and electrically sound.

The Lighting Designer leads the artistic side of the Lighting Team. The number of people who work with them is dependent on the size of the production. For most small- to mid-sized budget productions, the Lighting Designer works alone on their side of the team, but with larger budget projects, the Lighting Designer's team can grow significantly.

Underneath the Lighting Designer are Associate Lighting Designers and Assistant Lighting Designers. A production can have either one or the other or multiples of each and while the two positions are similar, they have some distinctions. Both roles manage production paperwork, track work notes, and have enough authority from the Lighting Designer to act as a surrogate when necessary. For example, if the Lighting Designer cannot attend a meeting, their Assistant or Associate may attend on their behalf. Similarly, it is common during the technical rehearsal process for the Lighting Supervisor to communicate with the Associate or Assistant about work notes or changes so that the Lighting Designer can continue work on another aspect of the project. The key distinction between Assistants and Associates is the degree of authority they have. Associates generally have enough artistic authority to make decisions or changes without consulting the Designer first, whereas the Assistants generally only execute specific instructions from the Designer. For example, an Associate may take a lighting design from one venue and rebuild it in another venue without the original designer's involvement, whereas an Assistant may only be permitted to do a redrafting or refocus of the same show from specific notes the Designer provides to them.

Next on the Lighting Designer's team is the Studio Assistant or Draftsperson. Some designers have enough work that they need to produce light plots quicker than they are able. For those designers, they will have an assistant that does not travel to theatres but rather receives rough, sometimes hand-sketched, plots from the Lighting Designer and then translates them into professional quality submissions.

Finally, the Lighting Designer may have an "Assistant to the Lighting Designer." This is a common role for an early career designer making the transition from college to professional life. Low paid or unpaid, these assistants help the designer with simple note taking tasks in exchange for the ability to learn through observation. An Assistant to the Lighting Designer differs from an Assistant Lighting Designer in that the Assistant to the Lighting Designer does not have the authority to speak on behalf of the Lighting Designer. In some cases, regional theatre will include Assistant to the Lighting Designer duties in their internships or apprenticeships to give those individuals exposure to both the technical and artistic sides of the team.

The technical side of the Lighting Team is led by the Lighting Supervisor or, in situations where there is no Lighting Supervisor, the Master Electrician. Some theatres employ Assistant and Associate Lighting Supervisors with a similar distinction to Assistant and Associate Lighting Designers. In general, these individuals share the macro-level department responsibilities of the Lighting Supervisor and usually have specific duties assigned to maintain the functionality of the department. For theatres with multiple venues, Assistants and Associates are usually given some autonomy either as "Master Electricians" on specific productions or as "Lighting Supervisors" of a particular venue.

As discussed, in certain contexts—particularly union contexts—it is common for there to be both a Lighting Supervisor and a Master Electrician or Head Electrician. In that situation, the Lighting Supervisor provides management support and supervision to the work of the Master Electrician or Head Electrician. The Master Electrician or Head Electrician are in turn responsible for the physical installation of the light plot and the management of the crew.

The crew is comprised of "Electricians." In an effort to avoid confusion with the electrical trades, these team members are now frequently called "Lighting Technicians." Lighting

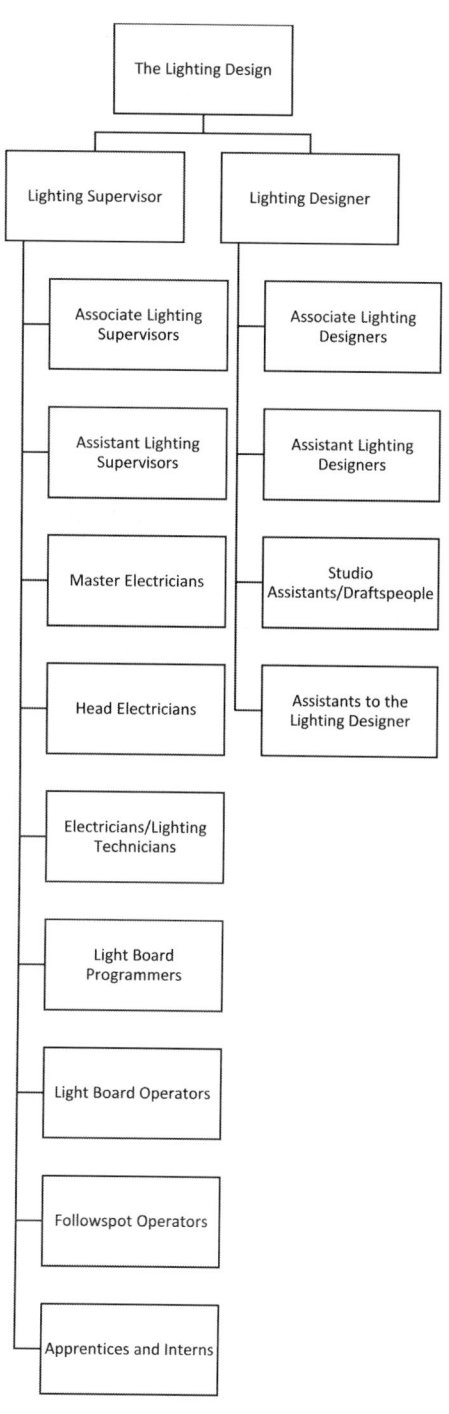

**Figure 1.5** The two sides of the Lighting Team.

Technician is largely the entry-level position for the technical side of the Lighting Team. However, many Technicians perform those duties for a number of years and in certain large venues are well-compensated for it. Consequently, the expectations of compensation and experience for a Lighting Technician are more tied to the company and region than anything else. Still, a Lighting Technician will always work under a more experienced leader and is therefore an ideal role for someone entering the industry.

Lighting Supervisors, Master Electricians, and Lighting Technicians comprise what is generally called the "Load-in Crew." They work in advance of the show to plan, prep, and install the lighting rig. However, when the technical rehearsals start, additional team members are added. The Light Board Programmer joins the team at this point to work directly with the Lighting Designer. The Programmer is a specialist with significant skill and experience on the lighting console that will be used on the production. Those in the top tier of Programmers generally freelance and travel extensively for their work. Although they report to the Lighting Supervisor, they work almost exclusively with the Lighting Designer and translate the general artistic needs of the designer into the language required by the lighting console. Their primary purpose is to allow the Lighting Designer to describe how they want the lights to evolve over the course of the production and have that be recorded in the lighting console without the Lighting Designer needing to know how that back-end process is done. For example, if the Lighting Designer says, "Let's take everything to zero and record a new cue with a time of two," the Programmer knows exactly how to translate that request into syntax for the lighting console in use.

Next are the "Run Crew." The Run Crew is comprised of the technicians in all disciplines that work during the performance itself to ensure that each performance is the same as the last. On the Lighting Team, the most common roles are the Light Board Operator and the Followspot Operator. The Light Board Operator takes over from the Light Board Programmer when the show is ready for an audience. The Light Board Operator makes sure that the lighting console plays back the show exactly how it was intended. In many small- and mid-sized theatres, the Light Board Programmer and the Light Board Operator are the same person and often not as skilled on the console as a Freelance Programmer. In these instances, the Lighting Supervisor should provide the Programmer/Operator with training to bolster those skills and Lighting Designers need to be more familiar with the syntax of the console being used. Additionally, in many cases the Light Board Operator is also responsible for the maintenance of the show after opening. They need to ensure that all the lights work each day and that their color and focus is the same as Opening Night. In large theatres—particularly union ones—these show maintenance duties are frequently performed by the Master Electrician or Head Electrician who then remains on-call at the venue during the performance, but for most small- and mid-sized organizations it is not practical to have a separate person for this maintenance.

The Followspot Operator is also a member of the Run Crew. Their job is to operate a specific light that needs to be focused tightly to an individual or group for certain moments in the play. In addition to operating their light, they are generally responsible for maintaining it during the full performance run. In small venues, the Followspot Operator is often an entry-level role, but for large productions it is not uncommon for the Followspot Operator to have many years of experience in that capacity. Followspotting takes minutes to learn and decades to master.

The roles of the Run Crew are all part of the Lighting Team and report to the Lighting Supervisor, however in most cases the Lighting Supervisor does not provide day-to-day oversight of these team members. That oversight is usually handled by the show's Stage Manager who shares direct supervision over all members of a show's Run Crew regardless of their discipline.

Finally, most theatres engage interns or apprentices. Historically, theatre training was all hands-on, very similar to the building trades. Even today, the best way to become a member of IATSE, the stagehands' union, is to join as an apprentice. However, union apprenticeships are different from the ones that are found at resident theatres in the United States. Most of the interns and apprentices in these resident theatres are post-college and are using the opportunity to learn how to translate their academic studies into a real-world job. Many of these apprenticeships or internships expose their apprentices or interns to both artistic and technical roles in the theatre so that they can explore which avenue suits their skills and interests.

## REFERENCE

Shelley, Steven Louis. 2009. *A Practical Guide to Stage Lighting*. 2nd Edition. Amsterdam. Elsevier, Inc.

# CHAPTER 2

# The Architect and the Engineer

## LIGHTING DESIGNER AS ARCHITECT

Lighting design for theatre is a complex beast, but it can be largely reduced to three artistic choices for each unit in the light plot: location, quality, and intensity. Intensity is how bright a light is at any point in the performance. The Lighting Designer must choose the available range of intensity through fixture selection as well as specific intensities throughout the show through cuing. Quality—the most complex choice—includes a series of interrelated attributes. The basic quality attributes are color, texture, and shape. Finally, location is simply where the light will be in relation to the actor or other object. When choosing location, the Lighting Designer is also determining the angle at which the light will hit the object.

Additionally, the Designer needs to evaluate how each choice impacts another. For example, if an arc source moving light is selected so that the Designer can have access to a higher intensity range, this choice will have an impact on quality because the lamp's color temperature will be notably cooler. Similarly, if a light is placed far away to create low-angled light from the front, the light will have less maximum intensity than if the same light was placed closer. Because each choice can so significantly impact the overall design, it is important that the Lighting Designer be able to make and communicate precise decisions.

It is equally important that the Lighting Supervisor is able to fully understand the choices made by the Lighting Designer as well as understand the priority of those choices. This understanding comes from both a study of Lighting Design and how it works but also communication with the Designer and their paperwork. Even before the design is submitted to the Lighting Supervisor, the Supervisor must be listening at design meetings and discussing the Designer's ideas one-on-one. The purpose of this time is never to lend one's suggestions to the Designer's process, but rather to learn what the Designer's process is. Perhaps the Designer is adamantly opposed to the color range of an LED

**Figure 2.1** *Fallingwater,* designed by architect Frank Lloyd Wright.
Library of Congress.

fixture or perhaps their aesthetic demands intensely bright light. Either way, understanding why the Designer makes the choices they make and how important each choice is to the overall design is crucial for the Lighting Supervisor to move the design forward to its final completion.

This relationship is analogous to the relationship between an Architect and an Engineer in building construction. An Architect must make many interrelated choices about the form and function of a building and an Engineer must be able to be in constant communication with them to understand how to support that design while also literally supporting the structure underneath it. The engineer must know how to negotiate the often-hidden underlying infrastructure that allow the art of the design to exist. For example, the famous Frank Lloyd Wright house, *Fallingwater*, pushed the engineering understanding of his day to the limit and since there was not a structural engineer equal to the aesthetic brilliance of Wright in 1937, 60 years later the design was beginning to fall apart. Luckily by 2001, engineering had caught up with Wright and Robert Silman, a structural engineer, was able to develop an infrastructure solution that would preserve the original design for decades to come (Wald 2001).

## LIGHTING SUPERVISOR AS ENGINEER

If the Lighting Designer is the Architect and the Lighting Supervisor is the Engineer, what is lighting engineering? Lighting engineering for theatre is comprised of five components: Structural Design, Electrical Design, Budgeting, Logistics, and Innovation. The Lighting Supervisor works within each of these components to maintain the authenticity of the Lighting Designer's artistic vision while ensuring a safe and seamless implementation of that design.

The first component, Structural Design, is an evaluation of the physical challenges of the design. In most theatres, designers are provided with a template of the venue that indicates standard hanging positions. However, these standard hanging positions are not always adequate to cover their desired fixture locations. Additionally, scenic elements could hinder the ability to use the standard hanging positions. It is up to the Lighting Supervisor to evaluate the Designer's specified fixture locations and determine what would need to happen—if anything—to allow the design to be implemented. What are the weight capacities of the positions? Do the pipes need to be counterbalanced and what are the capacities of the arbors? Do new hanging positions need to be installed to accommodate the design? Can the venue accommodate new positions?

The second component, Electrical Design, requires deeper analysis of the Designer's light plot. The imagination of a Designer is unlimited, but electricity and control architecture are full of limits. The Lighting Supervisor must evaluate each fixture, each lamp, each cable, and each dimmer to determine how to install the design within the limits of electricity and control. For example, the Designer might ask for a row of eight 750 watt

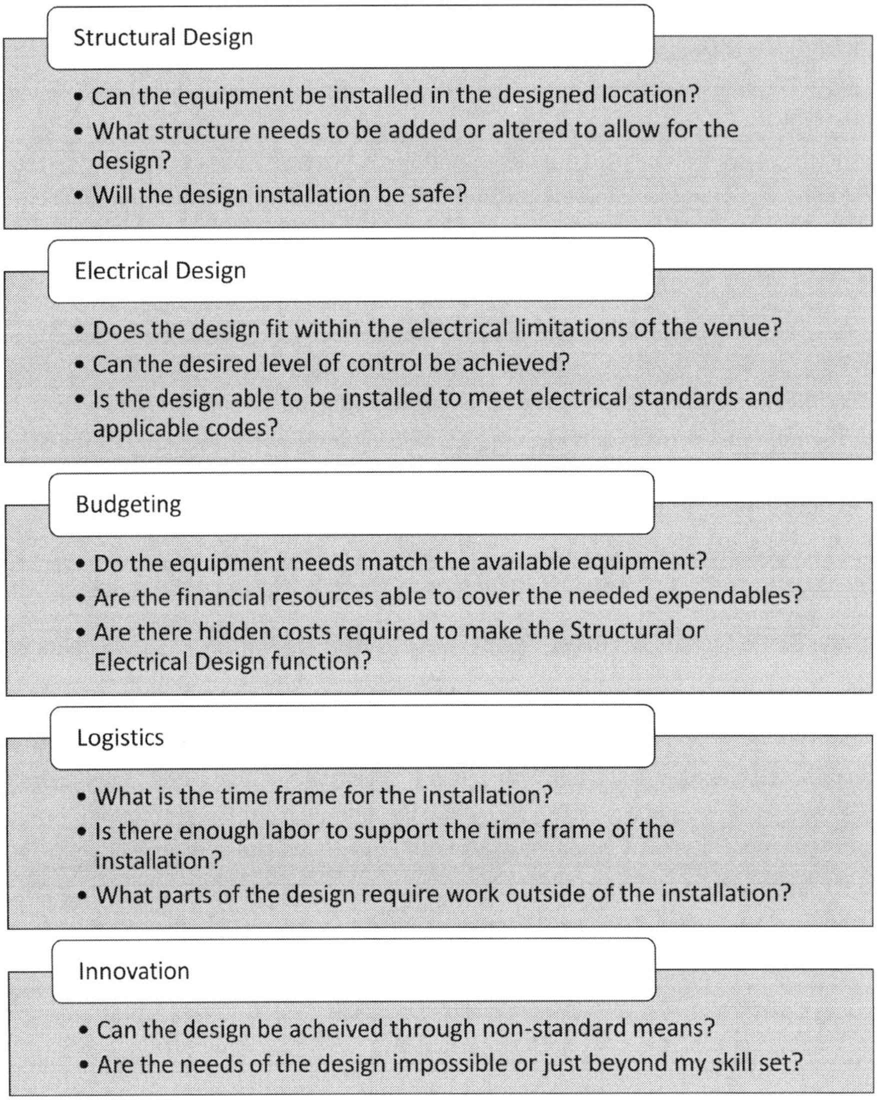

### Structural Design
- Can the equipment be installed in the designed location?
- What structure needs to be added or altered to allow for the design?
- Will the design installation be safe?

### Electrical Design
- Does the design fit within the electrical limitations of the venue?
- Can the desired level of control be achieved?
- Is the design able to be installed to meet electrical standards and applicable codes?

### Budgeting
- Do the equipment needs match the available equipment?
- Are the financial resources able to cover the needed expendables?
- Are there hidden costs required to make the Structural or Electrical Design function?

### Logistics
- What is the time frame for the installation?
- Is there enough labor to support the time frame of the installation?
- What parts of the design require work outside of the installation?

### Innovation
- Can the design be acheived through non-standard means?
- Are the needs of the design impossible or just beyond my skill set?

**Figure 2.2** The components of "lighting engineering."

lights in a row with the same channel. Can they be all circuited to the same dimmer? Must they be broken up somehow? Where will power come from? Another example might have a light plot with 40 LED fixtures each with a 15 DMX address range. What are the options to fit them on to one universe of control? Are there alternate modes that will reduce the

DMX range? How will control be distributed? Does the venue have adequate data infrastructure, or does it need to be supplemented?

The third component is Budgeting. For most companies that operate a venue, much of the equipment used is already owned by the venue. However, even in that situation there will be a need for specialty equipment, expendables, cross-departmental projects, and other supplies. When planning the structural and electrical design, the Lighting Supervisor is constantly evaluating the cost of the materials needed to execute their plan. If the cost is high, they will have to evaluate their plan for inefficiencies and then ultimately work with the Designer to recommend cuts that still preserve the essence of the design. A design might have 20 LED fixtures, for example, but the theatre does not own any. A purchase or rental of LEDs could be cost prohibitive, but the theatre does own color scrollers. Do the scrollers meet the fundamental needs of the design? If not, what is it about the LED fixture that is important to the Designer? How can that aesthetic be preserved?

Connected to Budgeting is the fourth component, Logistics. There is a finite window of time to install a production into the venue. The Lighting Supervisor must plan each day leading up to that installation window as well as each step of the installation. They need to determine how the design can fit into the available time and what labor needs are required to support that plan. For example, the Lighting Supervisor might make a plan that is one day too long but realize that one day of the installation can be done while the previous show is still in place. Alternatively, they may determine that hiring additional labor for each day will allow the installation to be finished one day sooner. Perhaps, neither

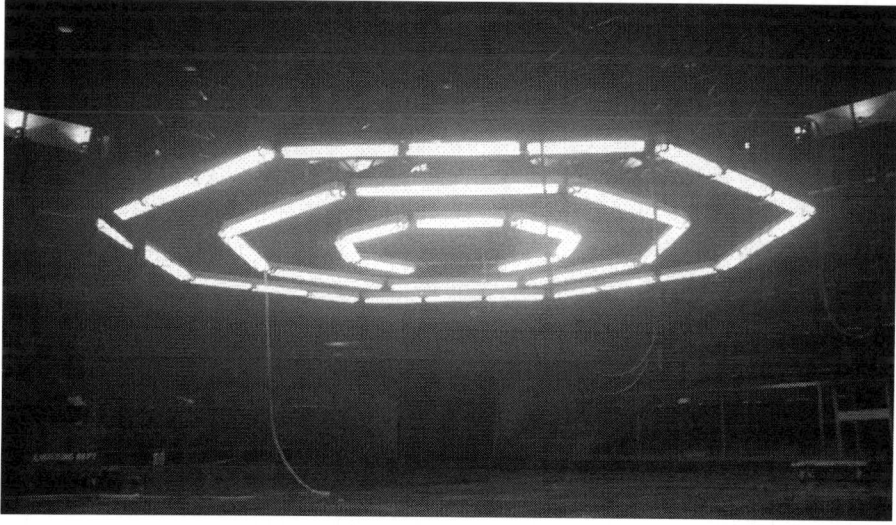

**Figure 2.3** To save money while preserving dimmability, the Lighting Team at Actors Theatre of Louisville built fake fluorescent tubes with LED tape and plastic guards.

are possible and then the Lighting Supervisor will need to work with the needs of the Designer to prioritize elements that may need to be cut or altered and thus reduce the installation time by one day.

The fifth and final component is Innovation. Innovation permeates the other four components. Often the equipment or processes the Lighting Supervisor has access to are not able to meet all the challenges of the design. In these cases, the Lighting Supervisor needs to determine if there is alternate equipment or other workarounds that can achieve the Designer's core artistic needs while still maintaining the structural, electrical, budgetary, and logistical soundness of the design. For example, if a central component of the design were a large fluorescent ceiling piece but dimmable fluorescents were cost prohibitive, the Lighting Supervisor could use existing resources to create the illusion of fluorescents with LED tape and plastic tubes or ask the Designer to forgo dimmability

## PEER TO PEER COLLABORATION

With the Lighting Designer as the Architect and the Lighting Supervisor as the Engineer, the most important parts of the collaboration are peership and respect of expertise. In many textbooks, the Lighting Designer is shown as the head of the Lighting Department and that the Master Electrician or Lighting Supervisor works at their direction. Consequently, many young Lighting Designers and Master Electricians operate within this ineffective power dynamic.

In professional settings, the reality is quite different. Here, the Lighting Designer and the Lighting Supervisor work in tandem as collaborators. A Lighting Designer cannot demand a design that is beyond the structural, electrical, budgetary, or logistical limitations of the producer, nor can a Lighting Supervisor artificially limit the design with disregard to the artistic needs of the production. Keeping these roles separate, yet in tight collaboration, provides a production with the ability to have expertise in both artistic and technical areas as well as a built-in checks and balances system. When this relationship exists in its' ideal form, the Lighting Designer can push the design to its artistic limits, knowing that the Lighting Supervisor will push back when necessary.

To make this collaboration work, both the Lighting Designer and the Lighting Supervisor need to be experts in their part of the process. Just as the Lighting Designer must know how a particular blue will make an actor appear, the Lighting Supervisor must know how much it will cost to maintain that gel for an eight-week run. These thought-processes work in concert as the work of the Lighting Supervisor is constantly supporting the artistic intent of the Lighting Designer and the Lighting Designer's work is constantly pivoting to fit within the engineering parameters set forth by the Lighting Supervisor. The two are in regular communication, both working toward the same goal—an excellent production that will be both artistically meaningful and functionally reliable.

In some situations, like educational settings or small professional organizations, the Lighting Designer and Lighting Supervisor are combined into the same role. This creates a unique challenge because the person in this combined role needs to work to provide their own technical limitations while not artificially limiting their artistic ambition. This scenario puts the production at a distinct disadvantage because it is unlikely that a single individual has the expertise to handle both the artistic and technical work at a high level. Additionally, it is a significant challenge to manage the attention to detail needed to be successful in either role. When I work in a combined capacity like this, I always subconsciously limit project complexity because I simply cannot handle the workload of a complex design without a collaborator.

It is easy to see this relationship as a battle. We imagine the Designer shouting, "But I need it!" and the Supervisor booming, "Well I said no!" This, too, would be an incorrect understanding. The Designer should not have to cry out, because the Lighting Supervisor should understand what is important. The Supervisor should not have to stomp their foot down because the Designer knows that the Supervisor researched and re-researched every option before concluding the element could not happen. If a battle erupts, then mutual respect is lost.

To add to the complexity of this relationship, the Lighting Designer is almost always a guest to the theatre while the Lighting Supervisor is almost always a resident within the company. When that is the case, the Lighting Supervisor is not only engineering a production, but rather engineering a whole season and beyond. The needs of one show may impair the needs of another show. This long-term outlook may be difficult for a Designer to envision without the context the Lighting Supervisor has. One Designer, for example, might want to paint all the lighting positions white, but the Lighting Supervisor knows there is not time to paint them back to black before the next show. A Lighting Designer will understand that this request is not possible within that context, but the Lighting Supervisor needs to share it.

## REFERENCE

Wald, Matthew L. (2001, September 2). Rescuing a world-famous but fragile house. *The New York Times*, pp. 1, 25.

# CHAPTER 3

# The Lean, Mean, Lighting Team

## LEADERSHIP IS EMPOWERMENT

Most Lighting Supervisors manage some combination of resident and temporary staff. Even if the Lighting Supervisor is also temporary, they will rarely work alone. This means that they will need to recruit, train, and supervise a team that fits the needs of the company, its' productions, or both.

Harry S. Truman's desk famously featured a placard reading, "The Buck Stops Here!" This meant that all the decisions of his administration were ultimately his responsibility (National Archives and Records Administration n.d.). The same is true for a Lighting Supervisor. All the technical work undertaken by the Lighting Department is their responsibility. Of course, in almost all situations, the amount of this work is so large that it is not possible for one person to do it all alone, so they engage a staff.

Not every company or production will be the same, so it is important to consider what sort of staff is needed to provide the support the production requires. In Chapter 1, there was a list of the possible roles of a Lighting Team. In a perfect world, every production would invest in highly focused experts assigned to each of these roles, but this is not always practical or necessary. For example, it is always great to have the console programmed by an experienced programmer who hands off the show to an experienced operator. However, dedicated programmers can be expensive and often beyond the capacity of small- and mid-sized producers. Practically, therefore, the Lighting Supervisor will need to consider the tasks of the production process with respect to the capacity of the producer and determine how some tasks can be combined to define a department's staffing structure.

When determining the staff needed for the Lighting Department, it is helpful to break the production process down into individual tasks and then assign those tasks to roles. In Figure 3.1, there is a list of typical tasks that a Lighting Department will undertake during a season. Some are production-specific tasks and others are general season tasks. In this example, the Lighting Supervisor takes on many of the general season tasks and hires a Staff Master Electrician to handle specific production planning and management. Then, they hire three Staff Lighting Technicians to provide installation and run crew labor. Followspotting tasks are left in a category for temporary or "overhire" staff. These staff members will only be hired if necessary and for only as long as they are needed.

While this example is not the only—or even the best—way to divide tasks, it helps to illustrate how the responsibilities of the department can be effectively delegated. To maximize efficiency, the Lighting Supervisor must be able to delegate the tasks of the department to their staff in a manner that highlights staff expertise. Here, it is acknowledged that the production planning and management work requires someone with specific "Master Electrician" expertise and a role is created with that expectation.

Some staff expertise may be discovered after hiring and roles may need to be tweaked to lean into a staff member's skill set. For example, it may turn out that one of the Lighting Technicians in the example is exceptionally skilled at console programming and operation. In that case, it could be more efficient to assign that person all the console responsibilities rather than have them be shared equally across Lighting Technicians. In this way, the Lighting Supervisor is honoring the demonstrated expertise of their staff member. When a staff member feels respected in this way, they are likely to do better work while simultaneously reducing the required oversight the supervisor needs to give them.

Of course, it is not always best to assign tasks exclusively on expertise. It can be equally effective to assign tasks based on a staff member's growth potential. In the above example, assigning all the programming tasks to only one of the three lighting technicians prevents the other two technicians from having an opportunity to grow that skill. To develop positive staff relationships, it is important to understand where each staff member is on their personal growth trajectory. If opportunities arise to assign tasks that can give someone a chance to explore a new skill, it allows the supervisor to demonstrate an interest in that person's growth. When staff feel challenged in their work, they have the potential to approach it with a greater intensity and attention to detail. Still, care must be taken in these situations because a staff member without expertise will likely need greater supervision to ensure accurate work.

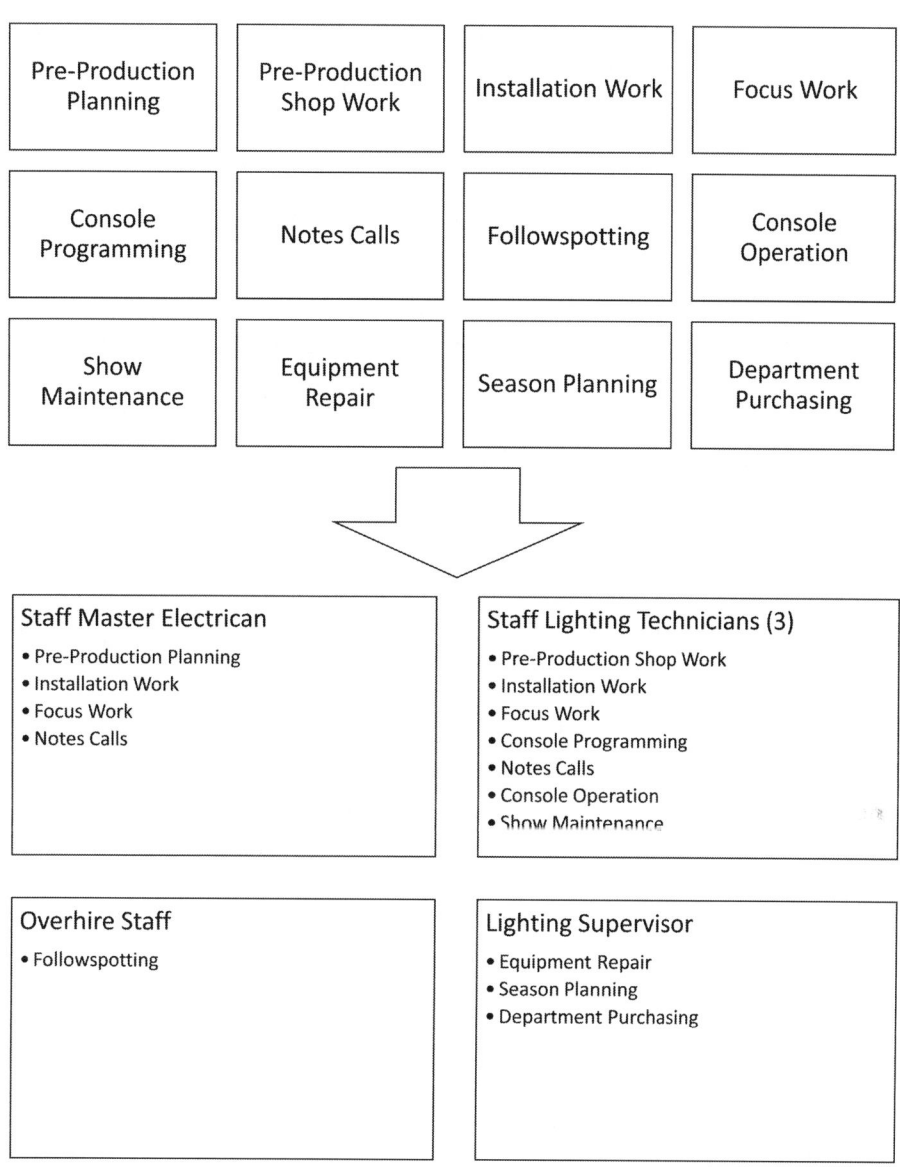

**Figure 3.1** Taking the tasks of the department and assigning them to the members of the department.

Balancing expertise and growth potential in the staff is essential in empowering them to work independently with a high attention to detail. It is easy for the "buck stops here" mentality to result in excessive oversight, but that would cause the Lighting Supervisor to disregard the expertise of their staff. Similarly, not allowing staff to develop new skills has the potential to make them feel undervalued and stuck in a certain hierarchical level. Consequently, a lack of empowerment could substantially impact productivity. As a leader, the Lighting Supervisor needs to ensure that staff get the opportunity to exercise their expertise as well as grow their skills. The resulting staff will be fully engaged in the work of the department and develop ownership over each production.

## DEFINING A ROLE

When hiring a position, it is important to define how the role fits into the structure of the department. Is the role a staff position that will work on multiple shows over the course of the season or will they work temporarily on one show or project? Will the role continue during any off-season down time when the company is not actively producing? What tasks will they be assigned and which of those tasks will need to be areas of expertise? Which areas can be learned on the job?

As shown in Figure 3.1, it is good to start by evaluating the tasks of the department and which roles will handle those tasks. Carefully consider these tasks as they do not all carry the same size or priority. For example, "Installation Work" and "Console Programming" are shown as the same size box, but installation is much more physically demanding than console programming. There will likely only need to be one programmer per show, while each production will almost certainly need more than one person working on the installation.

To visualize this, it may be better to represent "Installation Work" as multiple boxes, one for each required person. Figure 3.2 shows many of the same tasks as Figure 3.1, but here some tasks are repeated to underscore that they will have to be assigned to multiple people. In this example, each of the three technicians will be assigned the installation work, but only one will be assigned console programming. Therefore, it may be necessary to have the programming technician work under a different job description with different expertise requirements.

Similarly, some tasks may be low priority tasks that may not need to be assigned at all. In Figure 3.1, the Lighting Supervisor is handling "Equipment Repair" as part of their role. Equipment Repair is likely to be pushed to the back burner if the production calendar becomes particularly intense. Thus, it may be better to not include repair on the list of department tasks. Instead, it may be better to send equipment out for repair and consider it a material expense rather than a labor expense. With this change, the department would not need to have someone with repair expertise on hand.

Assigning these tasks to roles is the first step of developing job descriptions. Job descriptions are essential in identifying staffing needs and recruiting qualified candidates. The second part of developing a job description is to determine what sort of expertise

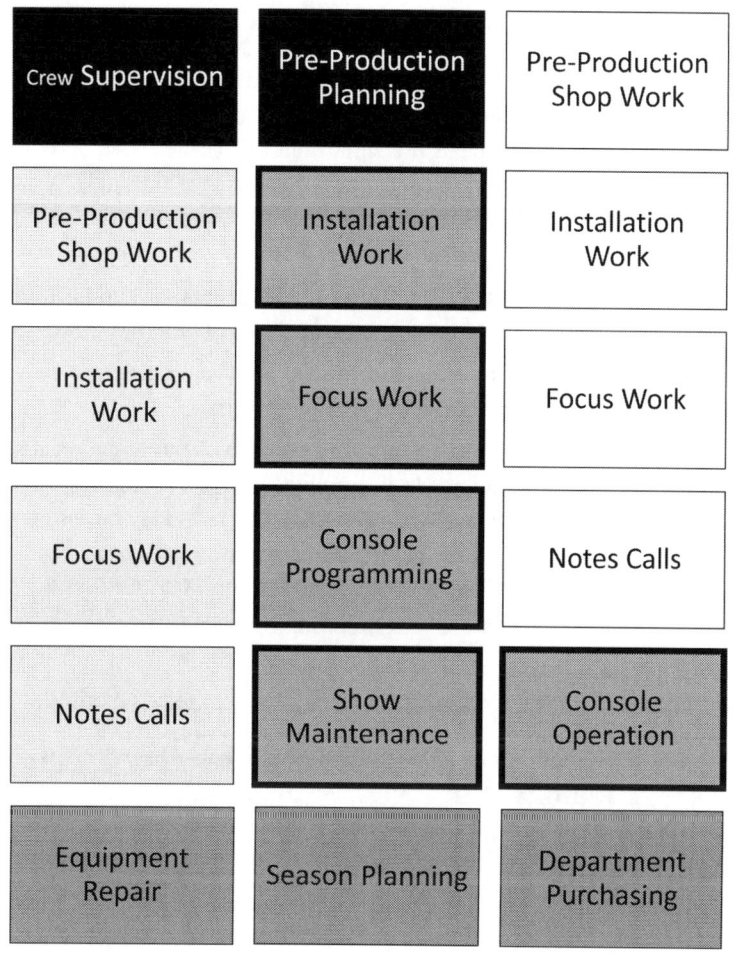

**Figure 3.2** Duplicate tasks that require multiple people to complete them.

will be needed to accomplish the listed tasks. In Figure 3.2, an example Staff Master Electrician's tasks are highlighted in black. To accomplish these tasks, they will need to have certain expertise like electrical and structural planning skills. When recruiting for this position, the Lighting Supervisor will need to consider how to advertise for these skills as well as evaluate each candidate to determine if they have the required expertise.

Developing roles in this way maintains a nod toward empowerment by developing a structure that avoids the unnecessary duplication of responsibilities, ensures adequate expertise, and gives staff a clear indication as what their role in the department's process is. Each person is an essential piece of the production puzzle.

# THREE HAT PHILOSOPHY

The process described in the last section works well for developing staffing structures for both temporary production teams as well as full-time Lighting Departments. However, when looking at full-time departments it is important to consider tasks beyond those of production such as asset management and annual maintenance.

When developing roles in a full-time context, it is important for each staff member to have a hand in both show tasks as well as in the general operation of the department or company. This larger role encourages investment in the organization and helps with staff empowerment. When developing these roles, I use a "Three Hat Philosophy."

Hat number one is the "Show Hat." Each team member is assigned show-specific responsibilities based on their expertise and growth potential.

The second hat is the "Department Hat." The specifics of this assignment should be tailored to each person's skill set and create a unique avenue for the person to contribute to the department or company. For example, a technician may be assigned weekly

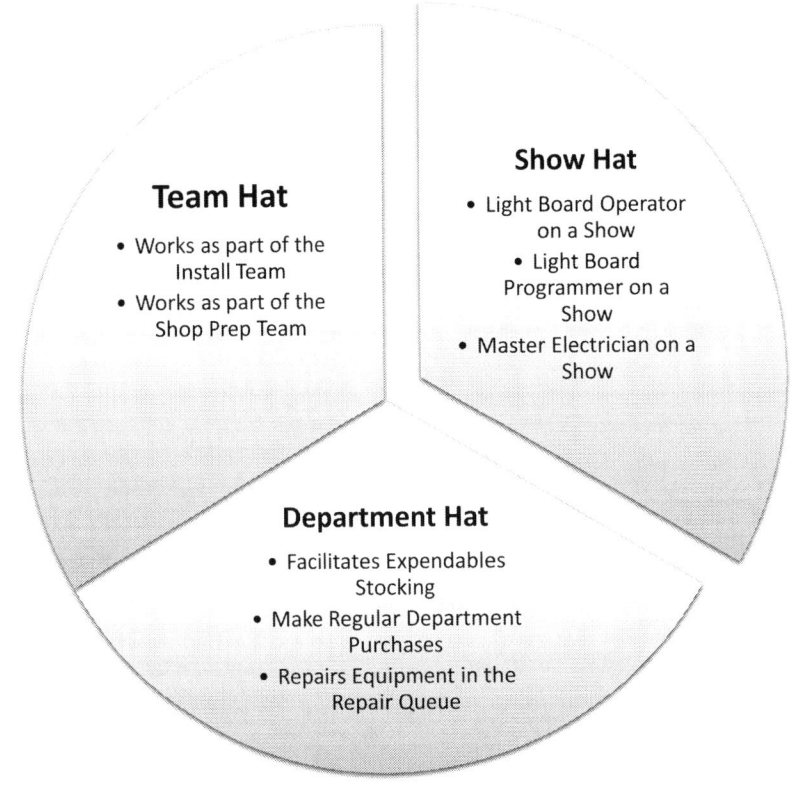

Figure 3.3 The "Three Hat Philosophy" of role development.

expendables counting and stock organization. This allows them to take charge of the storage of expendables and be the member of the department that knows the most about the current state of those needs. This second hat allows the technician to have a role in the long-term operation of the department and increases their ownership in the success of the organization at large.

The final hat is shared equally by all members of the team. This "Team Hat" is for all the tasks that everyone works on together to achieve the team's collective goals. Usually, these are ad hoc and relate to either the installation or preparation of shows. For example, all department members could be required to perform installation work regardless of their role. Assignments like these help to develop a team atmosphere where no staff member is above a certain base line work requirement.

## INDEPENDENT TOGETHER

Perhaps the most important component of role structure is the concept of "Independent Together." Henry Ford is famous for introducing the concept of the assembly line into our society, but the assembly line method is not appropriate for all workflows. The way a Lighting Team is most efficient is by working" independently together". Each member of the team has a specific task to do at any given point. They must be able to complete that task without interference from other members of the team, likewise they must not interfere in the process of their fellow teammates.

An assembly line method, for example, would have one person getting a light from storage, another transporting the light to the proper hang location, and a third hanging the light. Since each of these tasks are relatively simple, it is more efficient for each person to do all three tasks while working in different zones.

The concept of "independent together" is also very helpful in creating empowerment and equity. Each member of the team is responsible for making sure the work they do is accurate, and they cannot simply blame a teammate up the line for a mistake. Further, the total workload is spread evenly across the team so that no one technician feels they have done most of the difficult tasks. In fact, as every theatre has hanging positions that are more challenging than others, a Lighting Supervisor should take care when running installation crews to ensure that each technician has an equal amount challenging and less challenging positions to work on.

## GROUNDING

Maximum efficiency, however, does require certain situations where partner work is desired. The clearest example is in "Grounding." There may be times when the person hanging the lights is on a ladder or hydraulic lift and the storage of the lights is on the ground. In these situations, the person hanging the lights can increase their efficiency substantially by having someone on the ground hand them their lights. This person is called the "Grounder."

However, the key to a successful grounding relationship is an equal division of the total task. For example, when working independently hanging a light requires identifying which light is needed and where it must go, retrieving an appropriate light, and clamping it into the proper place. In a grounding relationship two people are working on these steps. The hanging technician focuses on clamping the lights into their proper place while the person on the ground identifies and retrieves the lights. Since the person on the ladder also needs to focus on their personal safety, the grounder should take a leadership role in the relationship and ensure that the hanging technician has all the information and equipment they need at all times. The grounder should always stay a few steps ahead of the hanging technician so that the hanging technician does not need to wait on them. The grounder stays ahead of the hanger by constantly knowing what the hanger will need and getting it ready for them. When the hanger is ready for their next light, the grounder hands it and the instructions off to them. The hanger simply listens to the instructions and hangs the light without any need to read the paperwork.

# REFERENCE

National Archives and Records Administration. n.d. "The Buck Stops Here Desk sign." *Harry S. Truman Library & Museum.* Accessed April 8, 2020. https://www.trumanlibrary.gov/education/trivia/buck-stops-here-sign.

CHAPTER 4

# Always Learning; Always Teaching

## A LEARNING EXPERT

Expertise is essential for any career. However, everyone is on a journey of expertise that is unlikely to ever be finished. A typical Lighting Supervisor will neither be at the beginning nor the end of such a journey. As you progress through your career, it is important that you command an honest expertise of the breadth of your work while simultaneously maintaining a humble approach to continued learning and alternate perspectives.

To be the "Lighting Engineer" discussed in earlier chapters, you must have a strong background in electrical practice and rigging. When considering your suitability for such a role you must be able to evaluate the breadth of your knowledge in these areas honestly. This may be difficult for early-career technicians to do as it is often hard to "know what you don't know." Typical prerequisites for a Lighting Supervisor are:

- Thorough understanding of the properties of electricity and associated math (Ohm's Law, etc.).
- Solid background in rigging practice for entertainment and related engineering concerns.
- Knowledge of all applicable codes and standards including an understanding of what work falls inside your scope and what work requires more qualified personnel.
- Broad knowledge of the standard equipment used in the industry and how it functions.

While this baseline certainly requires time and effort to develop, it is important for you to understand that when you reach this baseline there will still be blind spots in your expertise and further areas for you to grow into. Each scenario you encounter will likely have some component that is unique and if you enter these situations with complete confidence that you know everything you need to know, that hubris could result in dangerous situations. For example, if someone has spent their career in purpose-built theatre venues but now finds themselves in charge of an outdoor production, suddenly they are dealing with a whole host of infrastructure considerations that have always just been in place before. This does not mean they are completely unqualified to work in that situation, but it will require them to do additional research or engage additional support before proceeding.

Furthermore, as a Lighting Supervisor engages their team it is likely—and even desirable—that the Lighting Staff has expertise that is different than that of the Lighting Supervisor. A good manager of any industry will know how to use their team member's expertise to build the strength of the group at large. For example, if your team includes a freelance console programmer, it is likely that person will be more knowledgeable on the console than you. It is important to bring some humility when delegating that work. You need the programmer to feel confident in their own expertise so that the overall production can benefit. If you choose to be too prescriptive about how the programmer should work, the programmer will not feel comfortable enough to make choices in the heat of the moment without your input thus reducing their effectiveness.

Another component of this humble approach is that it allows the Lighting Supervisor to be open to further training opportunities. One of the great things about the Entertainment Technician Certification Program (E.T.C.P.) run by the Entertainment Services and Technology Association (E.S.T.A.) is that it requires those who have been certified to continue learning throughout their careers in order to maintain their certification.

It is easy for you to assume that once you become a Lighting Supervisor that you are done training. However, this cannot be further from the truth. Lighting technology grows at an exponential rate. Early in my career most Lighting Designers that I worked would go on at length about how LED technology would never be sufficient for professional productions. Today, I encounter many designers whom have built their entire aesthetic around the specific color qualities of LED lights. As a Lighting Supervisor, I need to be as versed in LED fixtures as I am in incandescent ones. If I had ignored those technology advances when they seemed like a long shot, I would have a lot of catching up to do today. You will always need to work to stay current.

The same approach is necessary for general working practice. Many Lighting Supervisors and other production department supervisors have ways of working that they see as ironclad, but it is important to keep your mind open. I have found that multiple department managers even within the same company can have conflicting "rules of thumb." In these situations, I like to hear the arguments on both sides and see how it compares to my experience. Almost always this helps me better understand how I should be working. You must remain open to learning new techniques and ideas whenever they come up. The world of theatre technology is huge. Constantly challenge what you know and constantly seek input from other professionals. Never work in a vacuum.

## LEADERS ARE TEACHERS

While it is important to honor the expertise of the team and remain humble, it is also true that most Lighting Supervisors find themselves with some degree of teaching responsibility. Whether your staff includes interns, entry-level technicians, or seasoned professionals, on-the-job training is an essential component of any role. A Lighting Supervisor needs to be able to develop techniques to encourage growth among their staff.

The best way to build a training relationship is to encourage the asking of questions. Staff that do not ask questions are more likely to do things improperly because they are not aware what properly is. To help encourage the asking of questions, the Lighting Supervisor should maintain a dialogue with each team member. Give them the opportunity to tell you how they work and why the make the choices that they do. If you do this, you will open the door for them to hear the same from you. Remember that when providing instruction, the "why" is just as important as the "what."

For example, a technician may be wrapping cable around a pipe rather than using tie line because that is what they did in school. You could simply say, "don't do that, you're wrong" but that shuts down the conversation. Alternatively, you could say "we like to run the cable under the pipe and tie it every 24 inches with tie line because that reduces the amount of time it takes to take it down later." Now the technician has a new method and a reason for doing it. When they encounter another practice from their school, they might take a moment to wonder if that practice is optimal and preemptively ask.

As another example, many early-career technicians hang lighting fixtures upside down. This is usually because they do not know there is a top in the first place. For them to understand not to hang a light upside down, they first must understand which side is the top and what makes that side the top. Once you start explaining—or better demonstrating—that the gobo slot is on top and the holder will fall out if hung upside down, it becomes easier for the technician to remember the rule.

Contextualizing is important especially when developing procedure. Your staff—particularly more experienced technicians—need to understand how a procedure came into being if they are going to internalize it. As a manager, the Lighting Supervisor needs to be able to justify their decision-making process because inevitably a staff member will challenge it. When that happens, it is key that you be able to explain how the established procedure or decision came to be.

For example, if during a changeover the policy is to strike all the cable runs, a technician might ask why they are striking a cable just to re-run it for the incoming show. In response to that question, you could explain why you concluded that striking and re-running was more efficient and—time permitting—work with the technician to evaluate the pros and cons of any alternate ideas. Even if the policy remains standing after the conversation, the technician will now understand the decision on a personal level and be able to take some ownership of it because they will have had the opportunity to reason it out for themselves. It is also entirely possible that a better way of working will come out of the discussion and that is equally positive. The important thing to remember in these exchanges is that you want to make the staff member feel as though they are part of the "we" in "this is how we do it."

## LEARNING RELATIONSHIPS

Many Lighting Departments are comprised of more than just a Lighting Supervisor and one other person. Usually there are multiple technicians or other staff. In larger departments like that there is an opportunity to encourage internal learning relationships.

For example, if a department has a Lighting Supervisor, an Assistant Lighting Supervisor, and two Lighting Technicians, the Assistant Lighting Supervisor needs to learn from the Lighting Supervisor but will also teach the Lighting Technicians. Additionally, the two Lighting Technicians will likely have different strengths and they will be able to learn from and teach each other as peers. In this dynamic, it is good for the Lighting Supervisor to encourage both the teaching from the Assistant Lighting Supervisor as well as the peer training from the Lighting Technicians.

This teaching becomes training in and of itself as it helps to prepare those staff members for leadership roles. It also helps to build confidence and team cohesion. It is easy for the Lighting Supervisor to always step in and provide the direction, but if the teaching is distributed there is greater benefit for the team at large. When someone shares their expertise with another person, they often develop a deeper understanding of it themselves. For example, I have had technicians train other technicians in day-to-day show running responsibilities. When they do this, they learn that they need to develop clarity and specificity in their own process so that to better convey it to someone else.

Finally, it is great team building to find time to reflect on the work the team is engaged in outside of the pressure of a production process. It is often hard to break down the merits of a procedure or idea in the heat of an installation. Scheduling time to have a post-production reflection—either one-on-one or as team—gives everyone the opportunity to celebrate achievements and endorse them going forward as well as work together to develop better methods for the next project.

# Part Two
# *Interrogation*

When I get a light plot from a designer, my first reaction is almost always, "What? Why? How?" "What is that moving light doing in the middle of this empty void?" "Why does this plot need double our inventory?" "How we will even *get* to *that* light?" Asking questions is my process. I never accept something at its' face value because there could always be hidden trouble behind it. I'm going to find that trouble. I'm going to head it off at the pass. I don't want to discover a problem in the middle of an installation with the time ticking away. I want to be prepared to answer the questions that will inevitably come from technicians. I want to solve the challenges of the plot so that when the audience sees the show, they will be the ones asking questions—"How did they do that?"

CHAPTER 5

# The Pre-Production Process

## ONBOARDING

The Lighting Designer is normally contracted by the Director of Production in coordination with the Artistic Department. Prior to that point, the Lighting Supervisor is not involved in the process. Once the designer is contracted, however, the Lighting Supervisor steps in as the Lighting Designer's primary producer contact. It is up to the Lighting Supervisor to start this relationship off on the right foot.

It is good to maintain a template designer introduction e-mail, as shown in Figure 5.1, to use for your first interaction. Ideally, this first e-mail will start an e-mail chain that can hold your entire pre-production conversation. It is best to keep this going rather than start new e-mails with new subjects. Keeping one e-mail chain allows the content of each conversation to be readily available and easily searchable in case of confusion. To facilitate a good starting place, your initial e-mail should contain a couple of key bits of information.

- Your name and contact information.
- A PDF of the venue inventory or link thereto.
- A Vectorworks or AutoCAD venue template drawing or link thereto.
- The plot deadline and what you expect to see on that date.

```
From: Jason Weber <jason@mytheatre.org>
To:   Lighting Designer <ld@ld.com>
Subject: PRODUCER NAME, PRODUCTION NAME - Lighting Introduction

LIGHTING DESIGNER'S NAME --

My name is Jason E. Weber and I am the Lighting Supervisor at PRODUCER NAME. I wanted to take a
moment to reach out to make sure you had the documentation and contact information you need as you
begin your process on our production of PRODUCTION NAME.

I have uploaded the available inventory and venue drawing to the Dropbox folder for this
production. Here is a link for reference: https://www.dropbox.com/xxxxxxxxxx. I'm also attaching
the inventory to this e-mail.

The inventory is two documents---an equipment count and a gobo catalog. We are always happy to
discuss the equipment and if you have ideas of things you need—whether they be house, rentals, or
purchases—the sooner you let us know the sooner we can start figuring that all out. Let us know if
you have any questions about the inventory.

I (he/him/his) will be your Master Electrician for this production and your contact throughout the
process. You can reach me via e-mail at jason@mytheatre.org or via cell phone at 123-456-7890

XXXX XXXXXX (she/her/hers), will be your programmer.

XXXX XXXXXX (they/them/theirs), will take over from XXXXXX after opening and run the show.

Your Light Plot is currently due on MONTH, DATE, YEAR, let me know if you foresee any issues with
that timeline. Again, if you have any thoughts about equipment or major set electrics as you begin
your thinking, please let us know, we'd love to start looking at pricing and planning!

If you have questions, let me know and I will be happy to sort things out to the best of my
abilities.

Cheers!

Jason E. Weber
Lighting Supervisor
```

**Figure 5.1** Template designer onboarding e-mail.

- Any oddities about the venue or process that the Designer should be aware of.
- A note to open the conversation around set electrics or practicals that the Designer may be aware of or to inform them about any conversations that you are aware of that they may not be.
- Names of other key lighting department contacts (staff assistant, programmer, operator, etc.).

The e-mail's subject should include the producer's name as well as the production name. Remember that the Lighting Designer could have ten or more active projects at a given time. It is important to give them a big searchable clue as to what production you are talking about. It is a good idea to include the producer 's name too in case they are working on the same production at multiple venues or—as sometimes is the case with new works—they forget the title of the play all together. It is also good to include the name of the producer and the production in the body of the e-mail as well to improve searchability.

When doing these e-mails, it is likely that the kind of e-mail sent to a designer you have worked with before is different than the kind of e-mail you would send to a designer that is new to you or your producer. Give yourself leeway to embellish on your template e-mail to develop a positive and professional first impression or to welcome back an old collaborator

It is possible that you or the designer will ultimately prefer to have conversations over telephone. While telephone conversations are great and help to provide some clarity that is difficult over e-mail, I recommend that you still keep an e-mail exchange active throughout the process. This way, you have a written place to go back to if there are any discrepancies about what was said. Additionally, if the designer is working on the plot at two in the morning and they need a quick reference about inventory or the plot deadline, it exists digitally.

The inventory attachment is likely the most important part of the e-mail. In some cases, the designer will read this attachment before reading the e-mail because this is the information they want. It is important that this information be presented concisely so that it will make sense to most readers without further explanation. The inventory should include:

- The available lighting console as well as its output capacity and software version.
- A description of the circuiting scheme of the theatre.
- A count of available dimmers and their capacities.
- A list of all instruments, accessories, and atmospheric equipment.
- A list of gobos and/or gel strings if that is applicable.

The bulk of the inventory is the list of equipment. It is good to be comprehensive with this list, but not all types of equipment are relevant to the designer so leaving off some of the irrelevant equipment is helpful for readability. For example, the designer does not care about your cable or power supply inventories, so those can be left off.

Additionally, grouping the equipment by category and subcategory will also help with readability. In the example shown in Figure 5.2, the "Ellipsoidals" are grouped together under the larger heading of "Instruments." This inventory has separate subcategories for "Wash Units," "Strip Lights," "Moving Lights," etc., which makes searching for the lights a designer needs quicker and easier. Note that despite this grouping, "Barrels" are listed with their corresponding fixture type because they are an option that is associated with that fixture. It is common to have more barrels or lens tubes than fixture bodies, so there should be an easy way to communicate that to the designer.

Finally, it is recommended that the counts provided to the designer be reduced by some percentage. This is your "percent spare." Inevitably, designers will want to use all the equipment you make available for them. When they do this, however, you do not have adequate resource should a fixture fail over the course of the production. This puts a strain on your team to fix the gear immediately or risk hurting the production. If you set aside a certain percentage—ten percent is standard—there will be equipment available to substitute in the event of an emergency. Most designers expect that this practice

### Designer Inventory Count

**Console**
ETC ION "Classic" (v. 2.9.1 - 4096 Outputs)

**Circuits**
192 x 20A in Grid; Mult Home Runs
6 x 50A in Grid; 2PG Home Runs

**Dimmers**
192 x ETC Sensor D20 @ 2400 watts
6 x ETC Sensor D50AF @ 6000 watts

*Note: If two wattages are listed, the first is the default and the second is an alternate. Please notify us if you wish to use the alternate.*

**Instrument**

*Elipsoidal*

| Manufacturer | Model | Venue Count |
|---|---|---|
| Altman | 1KL-6 40 Degree @ 1000w | 11 |
| Altman | 1KL-8 10 Degree @ 1000w | 7 |
| Altman | 3.5Q-5 (48-deg) @ 500w | 21 |
| Altman | 360Q 6x9 @ 750w | 27 |
|  | 360Q Barrel - 6x4.5 | 11 |
| ETC | Source Four Body @ 575/750w | 81 |
| ETC | Source Four Body (575 ONLY) @ 575w | 25 |
|  | Source Four Barrel - 05 Degree | 7 |
|  | Source Four Barrel - 14 Degree | 11 |
|  | Source Four Barrel - 19 Degree | 7 |
|  | Source Four Barrel - 26 Degree | 21 |
|  | Source Four Barrel - 36 Degree | 68 |
|  | Source Four Barrel - 50 Degree | 11 |
|  | Source Four Barrel - 70 Degree | 7 |
|  | Source Four Barrel - 90 Degree | 11 |
| ETC | Source Four Junior - 26 Degree @ 575w | 7 |
| ETC | Src Four Zoom (15-30) @ 575/750w | 21 |
| ETC | Src Four Zoom (25-50) @ 575/750w | 27 |

**Figure 5.2** Example designer inventory layout.

is happening and therefore will not concern themselves with these spare counts. It is up to the Lighting Supervisor to make sure the show is protected from equipment failures.

Of course, it is not always possible or logical to reserve spare. Small- and mid-sized companies frequently have only one or two of expensive line items like Moving Lights. In these situations, the Lighting Supervisor will have to determine how to mitigate a potential equipment failure without a spare on hand. They may research emergency rental options or increase the frequency of preventative maintenance. If you are running a fixture without a spare, discuss this with the Lighting Designer so that they know that it is possible that the gear could miss a performance. Discussing this possibility allows you to enlist them in creating mitigation options as well. Such options could be hanging additional conventional fixtures to cover important moving light usage or forgoing the use of such gear in show-critical applications.

## DESIGN MEETINGS

Normally, the onboarding process will be completed prior to the Initial Design Meeting. This meeting is the official beginning of the pre-production process, which lasts until the start of the build or rehearsal period. The pre-production process is when the design and concept for the production is created by the design team and the director in collaboration with the playwright. The role of the Lighting Supervisor in this period is largely observational. It is important that the technical team give the design team the ability to dream big during the design process. During this process, the technical team pays close attention to the design team's priorities. You want to know exactly what the design team imagines in a perfect world because there will be a time when you have to step in and ask them to cut things down to fit physics, budget, or other restraints. If you know and

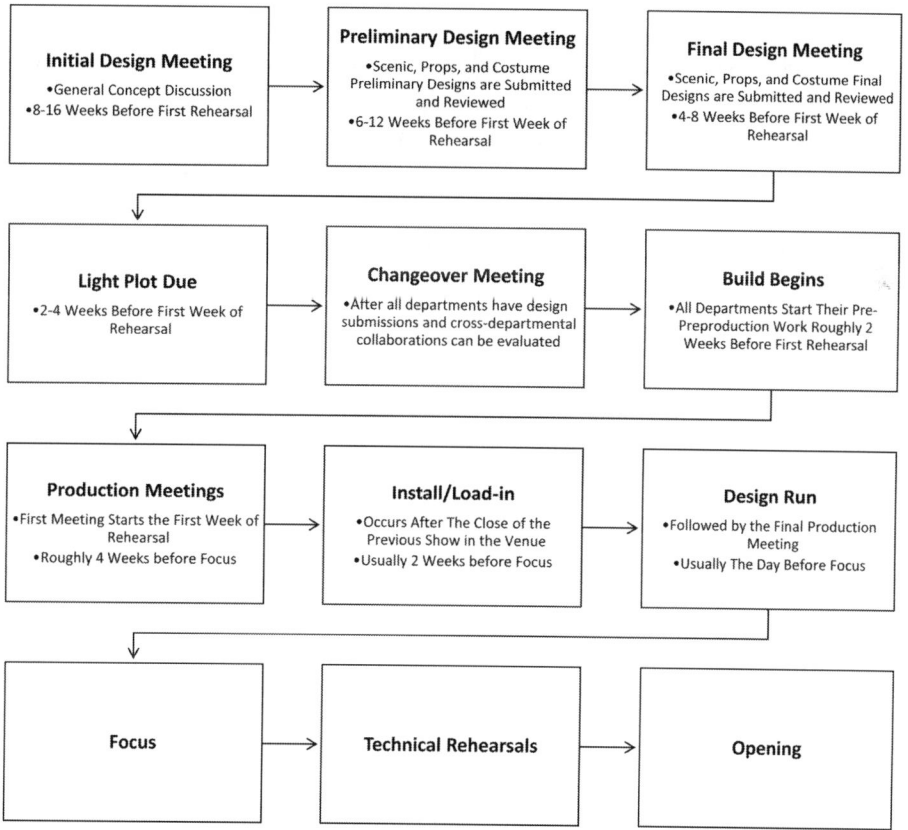

**Figure 5.3** Typical production process.

understand what the designer's dreams and priorities are, it is much easier to recommend logical cuts and adjustments.

As each meeting in the pre-production process moves closer to the real world the play will live in, the Lighting Supervisor should be evaluating the ideas discussed for cost and technical feasibility as soon as they come up. Engage with the designer in between meetings to understand what concerns or challenges they are facing. Sometimes designers limit themselves without your input. They may have an idea that they want to do but are worried it would be too outlandish. Try to get them to share those thoughts with you outside of the meetings so that you can evaluate them. It can be the case that there is a solution that allows for an idea that the designer does not see.

The Initial Design Meeting is the first and most theoretical of the design meetings. At that point, the design team, the director, and the playwright try to understand the script and how to translate it into a production. The Lighting Supervisor should make sure the Lighting Designer has their inventory and venue template prior to this meeting, so that they can understand the box that they are working in. Ideally, you and the designer would have talked through any quirks of the venue or the inventory so that as ideas get thrown about in this meeting, the designer is equipped to respond. It is common for the technical team to not participate in this meeting. If they attend, it is usually in a strictly observational capacity. Whether or not you attend the meeting, it is good to follow-up with the designer about any ideas that have arisen from it as these will likely be at the core of the final design.

Usually, the second design meeting is conducted as a response to the preliminary design submissions from the Scenic and Costume Designers. This meeting—The Preliminary Design Meeting or Post-Prelims Design Meeting—is the first time the design team will gather to discuss a concrete idea of the world of the production. Ideally, the design team and the technical team will have had the opportunity to review the preliminary design submissions prior to this meeting.

While there is not usually any submission from the Lighting Designer at the preliminary design stage, the Lighting Supervisor should still review all the submitted materials for any concerns or cross-departmental projects. If there are elements that would impact the Lighting Department, this is a good time to start a preliminary budget and installation schedule as discussed in the next chapter. Departments that receive preliminary design submissions for this meeting will start their budgeting process at this point, so it is important that they understand any cross-departmental impact. It may be that their schedule or materials projection is wrong because they did not consider any lighting conflicts. For example, the Technical Director might approve a two-week floor treatment that requires all the work to be done in the space because the installation window for the production is exactly two weeks long and it fits. They might not remember to consider that the Lighting Department needs to put ladders on stage to install the lighting equipment during that same two-week window.

Additionally, cross-departmental projects often use some budget allocation from multiple departments. For example, if the Preliminary Scenic Designs include a large marquee lighting effect, how much of that project will be covered by the Scenic Budget? It could be possible that the Scenic Budget can cover the entire cost including all the electrical expenses. However, it may be necessary for the electrical expenses to come out of the Lighting Budget. If a large scenic element like that eats up most or all of the Lighting Budget before the Lighting Designer has a chance to submit a plot, they can be severely limited. Sorting these cross-departmental issues out during the preliminary process gives the Lighting Designer a chance to negotiate with the rest of the design team before the design gets locked in. They may argue to the director that this money is better spent on a moving light rental package. If that is the case, the Scenic Designer will have to reimagine their design without the marquee.

Even if there are no issues or cross-department collaborations, discussing the submissions with the Lighting Designer is still a good idea. Find out what they are most excited about as well as what challenges they see. Be sure to listen to their thoughts and respond with any questions or concerns you might have. It is important to work with the designer in this preliminary part of the process so that the design does not get built around an unfeasible idea. Normally, the Lighting Designer does the speaking for the lighting perspective during design meetings. If they understand your technical concerns in advance, they can express them—and solutions—from an artistic point of view. It is easy for technicians to come across as "naysayers" in design meetings like this. Enlisting the Lighting Designer in these conversations allows logistical issues to be solved with creative solutions.

Finally, the scenic and costume designers will submit their final designs and the last meeting of the pre-production process—The Final Design Meeting or Post-Finals Design Meeting—will be held. At this point all the concerns associated with the preliminary submissions are addressed and incorporated into a final plan that the technical team will move forward with. In some processes, there will be multiple meetings between the Preliminary Design Meeting and the Final Design Meeting to shape the preliminary submission into the final submission. In other cases, these conversations will be conducted informally. In either situation, the Lighting Supervisor should remain in contact with the Lighting Designer and the other members of the technical team, so that they can respond to the design as it evolves. In an ideal world, when the teams gather for the final design meeting, everyone will have seen the designs already and everyone will have had their concerns addressed. In that way, the purpose of the meeting is primarily to make sure the whole team is on the same page before the production process begins in earnest.

## THE PLOT SUBMISSION

It is usually the case that the lighting design submission—the light plot and associated paperwork—is submitted after the Final Design Meeting. This can seem counterintuitive, but it is because the lighting design needs to be coordinated with and respond to the scenic and costume designs. Because of this later submission deadline, it is crucial for both the Lighting Designer and the Lighting Supervisor to be actively involved in the design meeting process. The design meetings are the opportunity for the lighting team to create the avenue that the lighting design will occupy within the larger design concept.

The deadline for the plot submission will be set prior to the beginning of the pre-production process as it must be included in the Lighting Designer's contract. The Lighting Supervisor should collaborate with the Director of Production and the other members of the technical team to develop the deadline and meeting schedule for the entire pre-production process leading up to and including the plot submission deadline. There are three things the Lighting Supervisor should consider when setting this deadline.

- How much time does the lighting team need to plan and install the lighting rig prior to focus?
- How much time does the Lighting Designer need to create the light plot after the Final Design Meeting?
- When is the Lighting Supervisor able to review and respond to the plot submission?

How much time the lighting team will need is going to be unique for each producer and venue. Four weeks prior to focus is a safe rule of thumb. That will allow two weeks for planning and two weeks for installation. However, if you have multiple venues and overlapping schedules you may need more time as you might have to hit the pause button on one venue to work in another. If that is the case, it might make sense to set the deadline five or six weeks prior to focus. In fast-paced environments like summer theatres, staff might not be contracted four weeks prior to focus so you may have to make do with a two-week period. Whatever your situation, it is important to consider how long it will take you to plan the installation as well as how long you have to complete the installation. If the plot is due the day you are supposed to be installing in the theatre, you will not be able to start right away.

It is also important to note that in certain circumstances the time you need is not equal to the time you have available. When that is the case it is important to discuss this limitation with the Lighting Designer at the very beginning of the process so that they understand that the scope of the design will have to tempered to accommodate this scarcity in planning or installation time. If you wait to share that limitation with the designer until after the plot submission, you risk having even less time while you wait for a revised plot.

Second, the deadline must take into consideration the amount of time the Lighting Designer needs to make the plot. This is important because if the designer is not able to do their best work, the plot that you ultimately receive will be incomplete and require revision thus delaying your ability to begin the planning process. I recommend that a

Lighting Designer have two to four weeks between the final scenic design submission and their plot deadline whenever possible.

The final consideration is the Lighting Supervisor's schedule. Most professional Lighting Designers work on many shows at once so when they submit their plot to you, they will likely move on to their next deadline. It is essential, then, for you to be able to get back to them with questions, concerns, or other feedback within 24 hours of the submission deadline. Schedule the deadline for a time that you can dedicate at least a full eight-hour day to review and do your best to protect that day in your calendar. To help with preserving your review day, clarify with your designer in the onboarding period what time on the deadline day you expect to receive the plot. It can be frustrating if you schedule a full day to review the plot and the plot comes in at 11:59 p.m. on that day.

In addition to clarifying the deadline, it can be helpful to clarify what the design submission package is expected to contain. Most designers understand these expectations through years of experience but would not be offended with a casual description of your needs when you send them their onboarding message.

At a minimum the lighting design submission should include.

- A scale Light Plot—ideally in PDF format as well as in AutoCAD or Vectorworks format.
- An Instrument Schedule and Channel Hook-up—ideally in PDF format as well as in Lightwright format.

Many designers will also provide other helpful documentation such as equipment lists or shop orders. If these are helpful to your process, there is no problem in asking for them.

Once the plot is submitted, the final step of the pre-production process is the Changeover Meeting or Installation Meeting. This is a meeting between the members of the technical teams. In this meeting they work through much of the cross-departmental issues they may have. At a minimum, this will include scheduling the time each department will be in the venue during the installation so that one department does not unnecessarily hinder another. For example, if the Scenic Department wants to remove the floor on day one, but the Lighting Department needs to set-up a ladder on the same day that could be a problem. Similarly, the way departments will need to collaborate on cross-departmental projects like practicals and set electrics should be established in this meeting. With all these plans in place each department can move forward with preparing, building, and installing their designs.

CHAPTER 6

# The Review and the Price Out

## PLOT REVIEW

After the pre-production phase, the next phase is the Prep Process. This is where the Lighting Supervisor does their most important work. An effective Prep Process will set-up a smooth installation and ensure that the design functions effectively during technical rehearsals and performance. This process—also called "advancing a show"—has four distinct components: Plot Review, Price Out, Paperwork Prep, and Shop Prep. The Prep Process is crucial to the success of the show because once the clock starts on installation, delays can be exponentially detrimental. The Prep Process gives the Lighting Supervisor the opportunity to plan out the installation and preempt any issues before the team gets in to the theatre.

The first component of the Prep Process is Plot Review. Plot Review begins as soon as the Lighting Designer's submission arrives. The goal of the Plot Review is to determine if there are any concerns, questions, or issues with the design before moving forward with deeper planning. You want to make sure that you can catch issues early so that they can be resolved before you get too far along with your prep work.

## SETTING UP YOUR REVIEW NOTES

You can take your review notes on paper or digitally. I normally keep it simple and use a note pad. Divide the notepad into two sections—"Red Flags" and "Yellow Flags." Red Flags are notes that you want to discuss with the Lighting Designer right away. Mostly these are problems that keep the design from fitting within the budget or insurmountable technical challenges. Sometimes they can also be irregularities that you find in the designer's paperwork. Yellow Flags are primarily notes to yourself so that you remember to look deeper at certain areas. In the end, Yellow Flags might become Red Flags or they might go away altogether.

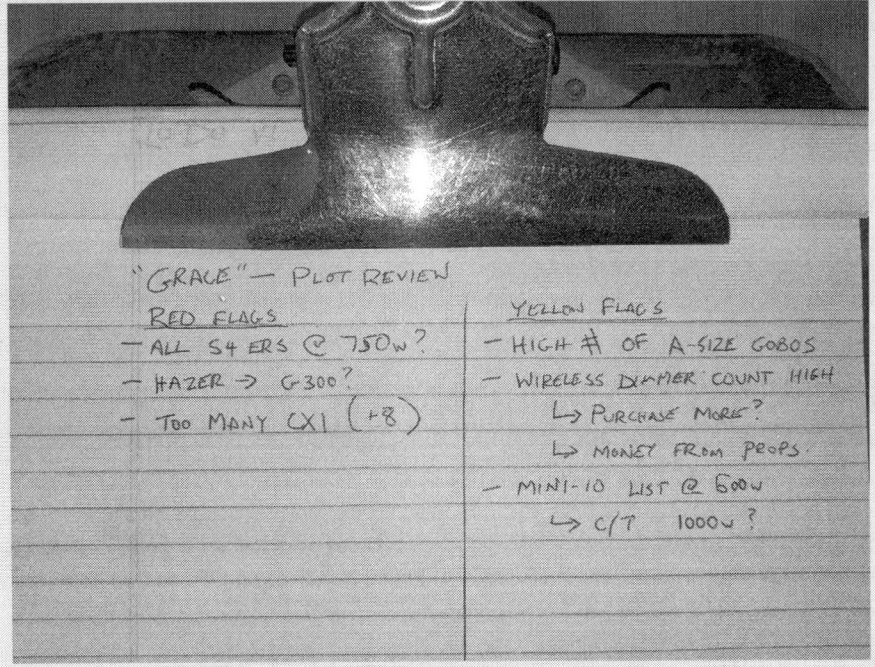

**Figure 6.1** Review notes sheet.

To begin the Plot Review, the Lighting Supervisor scans the drawing and the submitted paperwork for any red flags. The scanning process is aided by experience as certain patterns in designer's choices emerge over time allowing irregularities to jump off the page. Early-career Lighting Supervisors will want to take extra care to do this process methodically so that they can see these patterns.

When a designer makes a choice that goes against a typical pattern, you should flag it as a question. For example, Rosco 119 is a diffusion filter and Lee 119 is a dark blue color filter. It is common to see Rosco 119 paired with another color filter in the same fixture. However, it would be less likely—although not impossible—that Lee 119 would be paired with another color filter. So, if you saw an entry that read "R08+L119" suggesting that Rosco 08 and Lee 119 would both be assigned to the same fixture, you would recognize that as a potential red flag since it goes against typical pattern.

It is important to note that, in this example, while you might be nearly certain that the designer meant Rosco 119, you should still ask it as a question before making the change. At least one time out of a hundred something that you think is wrong will be right. This is why it is better to think of these issues as "irregularities" rather than "mistakes." You do not want to simply assume the designer made a mistake and correct it. If you do that regularly, you will eventually find yourself at a focus call with an angry designer.

Of course, it is just as important to not assume that these irregularities are intentional. You do not want to spend time planning out an elaborate solution for a challenging problem that turns out to be a misunderstanding or typo. For example, if the Lighting Designer draws a floor mount inside the audience risers, it is unlikely that they intended for the audience riser to be excavated to accommodate the light. More likely, they did not realize the obstruction existed in the first place. If you simply red flag that irregularity and let them know that light cannot penetrate the seating risers, you can both have a good laugh and make a new plan.

During this initial pass, you should keep a running list of these red flags so that you can ask about them in one e-mail or phone call. It is always best to streamline the number of times you contact the Lighting Designer. The more times you contact them the more likely a message will get lost or a question go unanswered. Again, e-mail is recommended for these exchanges so that you both can maintain a record of questions and responses. Ideally, each of these e-mails are attached to the same thread that you started with the initial onboarding e-mail so that the full conversation can be easily found and searched. However, if that thread has been broken and a new one needs to be started that is not an issue. If that happens, be sure to include the production name and the producer name in the new e-mail's subject line just like you did in the onboarding e-mail.

After this initial pass, if there are any serious issues send off your red flag e-mail right away so that you can get a corrected plot before continuing. If there are no issues or if the issues are minor, you can move on to your second pass. This second pass is a more in-depth review that allows you to look at each fixture with greater scrutiny. The goal here is to find things that present technical, budgetary, or logistical challenges. In this second pass, you may add to the list of red flags or start a list of yellow flags.

Your yellow flags are things that will require you to find a solution for but are not inherently impossible. For example, booms and added positions are always yellow flags. In these instances, the designer is asking that lights be installed on hanging positions that do not currently exist. As the engineer, you are going to have to construct new hanging positions which will require a review of materials as well as available rigging points. Flagging these elements highlights that you are going to have to go back and make those plans. Other possible yellow flags include fixture counts that are over the designer's inventory, set electrics or practicals that were not on the scenic drawings, or elements that require additional crew such as followspots. As you move through the budgeting and planning process, you may discover that yellow flags become red flags and need designer attention, but at this point you are primarily keeping them as notes for yourself.

## PLOT CLEAN-UP

To make this second pass, I use a "plot conversion" or "plot clean-up" process. This is not to say that all lighting design submissions need to be "cleaned-up" because they are messy or poorly done, but just as a Technical Director creates build drawings for the shop and rarely builds from the Scenic Designer's submission, the Lighting Supervisor needs to transition the Lighting Designer's submission to a "hang plot" and "hang paperwork." When you do this for the first time it can seem inefficient, but the "plot clean-up" process has three distinct advantages.

- It forces the Lighting Supervisor to look at all the details of each fixture, position, and other special projects for red or yellow flags.
- It allows the paperwork to be reformatted to match a house standard that staff can easily recognize.
- Preserves the designer's initial submission as a reference while allowing the Lighting Supervisor latitude to add power and data planning information to the paperwork without losing that reference.

To begin the "clean-up," create a folder on your computer that you will call "from LD." Put all the files the designer has sent you in that folder. If you want, you can make subfolders with the date of each submission. This is helpful if you start to receive multiple submissions or revisions, but either way the goal is to put aside the submitted files to protect them from accidental editing. Then, save a copy of the AutoCAD or Vectorworks files and the Lightwright File in a separate folder. These will be the files that you will "clean-up," so this folder should be the one you intend to work out of.

There are many ways to go about the clean-up process and undoubtedly you will develop your own method that works best for you. Here I describe the method that I use, but that is not intended to suggest that it is the only way to do this work. I encourage you to use this workflow as a jumping off point to find your own path. Further, I work exclusively in Vectorworks and Lightwright. Other Lighting Supervisors may use AutoCAD instead of Vectorworks or another spreadsheet or database program instead of Lightwright. I have found that the Vectorworks/Lightwright combination is extremely common among lighting professionals, so even if it is not the software combination that you currently use, I would recommend considering it.

As a starting place, I keep a Lightwright Template file and a Vectorworks Template file for each of the venues I work in. The Vectorworks Template file is the one I sent to the designer during the onboarding process. In addition to this being a template venue drawing, I include a symbol library, which conveniently has symbols for each of the fixtures and accessories that are in the inventory for that venue.

I do not normally send my template Lightwright File to the designer because it is not likely that they would use it. When generating their Lightwright File, Lighting Designers generally prefer to use their own template set-up. Since this means that my Lightwright File is set up just for my use, I can be very specific with how I do it. The areas that I normally configure are:

- LW Vocabulary Options so that text is expressed in the way that I expect it to be.
- Maintenance for Instruments, Positions, Accessories, and Color Frames so that they are prepopulated with data related to the venue and the producing company.
- Dimming and Control so that my venue is accurately represented.
- Worksheet Column layouts for each sort option.
- Print layouts for each paperwork type that I normally use.
- Worksheet Line Items for the House Lights or other Utility Lighting.

Since my template file and the designer's Lightwright File are separated, I need to use the "Merge Show File" function in Lightwright to combine them. To begin, I copy my template Lightwright File into the same folder as my "clean-up" files and rename it as if it was my new show Lightwright File. For example, if it were "LW-Template.lw6," I would rename it to "Show Name – LW – Date.lw6" or something that clearly designates it as the active Lightwright File. To avoid confusion, I would then rename the designer's Lightwright File to something like "Show Name – LW – Current.lw6."

With the two Lightwright Files in the same folder, I merge all the worksheet data from the designer's file into the template file. I make sure to not copy any other maintenance or set-up fields because I want to keep the set-up from my template file and do not want the designer's file to overwrite it. As I do this, I pay special attention to the fields from the designer's file that have data in them. This is an opportunity to look for any unexpected fields that are being used. For example, some designers use a "user column" for certain items like diffusion or iron. Without this check, you may miss that there are data in those columns. If you see data in those columns, be sure to make a yellow flag to look at it and move it to fields that make sense with your workflow. When you execute this merge, the software will likely give you a warning because the show files started as quite different files. This is expected, so proceed.

When completed, you will have one Lightwright File that has all the data for the show and is set-up how you like. At this point, you are done with the designer's file unless you

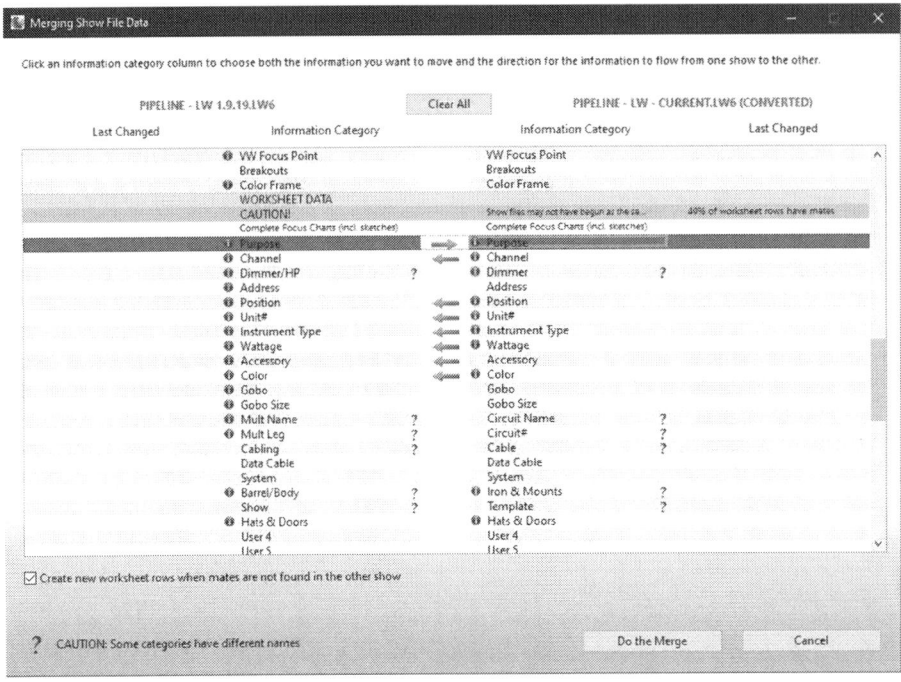

**Figure 6.2** Lightwright's "Merge Show File" dialogue.

need it for reference, so I usually delete it or move it back to the "from LD" folder if there is not a copy of it in there already.

With the data merged, it is time to inspect the new Lightwright File. I do this by reviewing each "Maintenance" screen and looking for entries that did not pair with the entries I set-up within the template file. This will happen when the designer used entries that were listed differently from what the template file has or when the designer has used instrument types that the venue does not have. In the case of the former, this gives me an opportunity to conform the entries to what the technicians expect to see and in the case of the later, it gives me a yellow flag that a rental may be needed for the show.

Conforming the entries to match existing entries is easy. Simply, change the text in the left most column entry to match that of an existing entry. The software will ask if you want to combine entries, and you will say that you do. For example, the designer's entry might say, "Source4-36deg," but I prefer, "ETC S4-36." When I change "Source4-36deg" to ETC S4-36" in the maintenance window, Lightwright will ask if it should combine. When I say that it should, it will apply all my template data—color frame, weight, symbol, etc.—to the fixtures in the show.

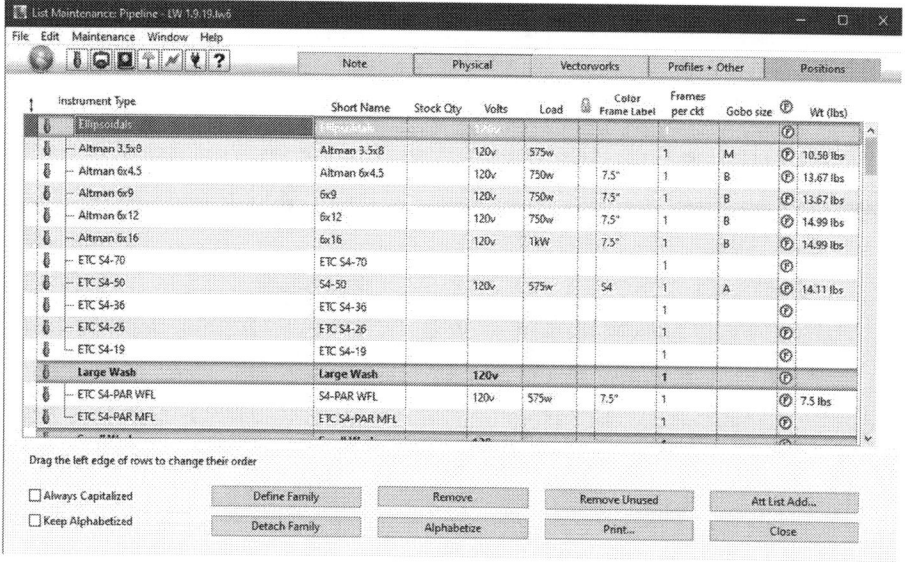

**Figure 6.3** Editing Lightwright's "Instrument Maintenance".

After I have done this with all the maintenance dialogues, I turn my attention to the Vectorworks file. I will use the Vectorworks link feature to combine my newly cleaned Lightwright File with the designer's Vectorworks File. Since I have changed the instrument type names to make my default symbol names, I will have to make sure that the Vectorworks Files that the Lighting Designer submitted still contains my symbols. Since most designers have their own symbol libraries that they like to work with, it is common that they will delete any of your symbols. Luckily, Vectorworks has an easy function in the Resource Browser to import symbols from another file. So, before I enable the Lightwright-Vectorworks link, I will check the Vectorworks File to ensure that it has my template symbols, label legends, or titleblocks and, if not, import them. When I know the Vectorworks File is ready for the merge, I rename it to match my active file naming convention: "Show Name – VW – Date.vwx."

Once the file is renamed and has the proper references inside of it, I follow the link instructions in Lightwright. When prompted, I indicate that all the fields in Lightwright are correct and should overwrite the fields in Vectorworks. This will make all the fields in Vectorworks match the Lightwright template names. Additionally, it will convert all the symbols in the plot into the template symbols that you just imported.

While I normally do a good amount more in the Vectorworks File to ensure that I am acquainted with the layer and class structure that the designer was using—sometimes I

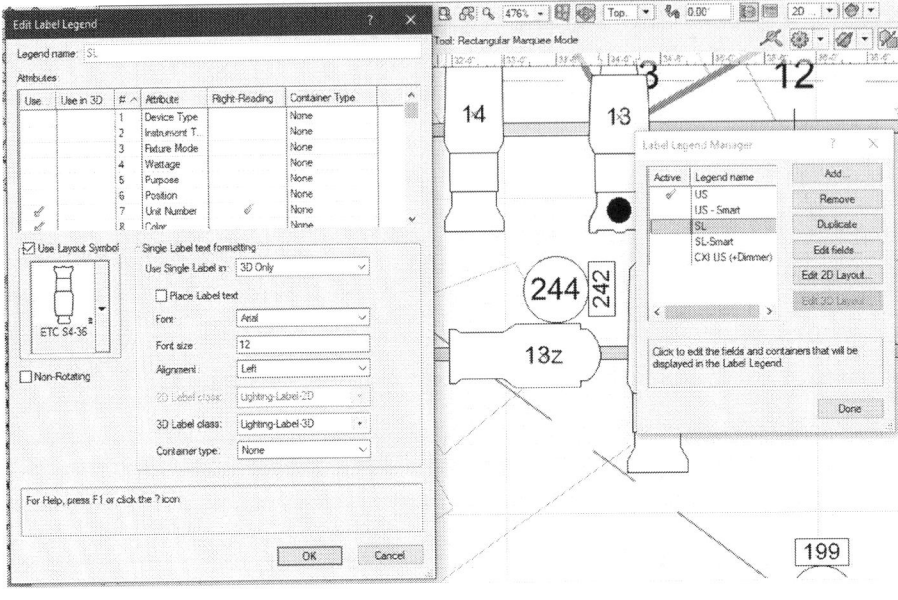

**Figure 6.4** Vectorworks' "Label Legends Manager".

modify it to make it simpler for me to use, other times I just learn it and move on—the next step is the last step of my official clean-up process: Label Legends.

When I checked the Vectorworks File for symbols, I also made sure to import my Label Legends. Label Legends are the method Vectorworks uses to display data for each symbol. Most designers will have a set that they have used to create the plot, but those often exclude valuable power and data information. Once I have imported my Label Legends, I will select all the units—either in Lightwright or Vectorworks—and change the Label Legend to one of mine. With all the units reassigned, I will delete the Lighting Designer's Label Legends to avoid confusion.

This process will assign the same Label Legend to all the fixtures in the light plot which is not necessarily ideal, so I next need to look at each light in the light plot to determine which Label Legend is best. Because this process requires me to look at each light in the plot and the information assigned to it, I can note any yellow or red flags that come up. As an example, I may set a Label Legend for a light and then notice that it has the same unit number as the light next to it. If a light has the same position and unit number in Lightwright, the system will treat it as the same light for counting purposes and that will prevent me from knowing the true count. I may even discover that the light in question has position information that conflicts with where it is drawn. Both things are potential flags to note.

With the file cleaned-up, you are ready to pass all your questions on to the designer and move on to the next component of the Prep Process—Price Out.

## PRICE OUT

Pricing out a show is the translation of the design into resources. Here, you think through each element of the design and determine what is needed to complete it. For each element of the design, it is important to price out labor, materials, and time. Time is the least malleable, so it is best to start there.

To price out for time, make a schedule with a goal for each day of the process starting with the current day and ending with focus. Your goal is to have the design fully realized and installed prior to focus. The importance of this cannot be understated. The plot you install will almost certainly have changes that occur prior to or during the tech process. These changes most often begin to appear during the focus call and continue until the opening of the show. Any of the initial installation work that is not completed prior to focus will hinder your team's ability to handle these changes, so always plan to start focus with a blank "to do" list.

When making your schedule, it is important to think about the goals of the Prep Process itself. An efficient installation is contingent on an effective Prep Process. If you do not provide time to prep for your installation, that prep work will inevitably happen during it. For example, it is common to cut and frame color filters prior to the start of installation. That process itself could take a whole day if the show is large enough. If you planned for an hour of installation to be for dropping color but did not leave any time to cut and frame color, your schedule could easily fall a day behind.

For the schedule itself, I use a simple spreadsheet, as shown in Figure 6.5, that shows a line for each day. Normally, I break my goals into morning goals and afternoon goals. This is not necessary, but this level of granularity helps my process. Your process may be helped with greater granularity. For example, you may have four sections for each day so that at each rest break you can check in on your progress. Conversely, you may find that every segment of your process takes a whole day and that there is little value in setting subgoals. Whatever your needs are, I recommend dividing into the maximum amount of granularity that you can wrap your head around. The more detail you put into the schedule, the more realistic it is. The ultimate goal of making the schedule is to determine if the project can be completed in the time allowed.

In addition to my goals, I make an external conflicts column that helps me track what other departments are working on. Often in this first pass this column is blank. When I take my schedule to design meetings or the changeover meeting, I fill in with conflicts that get discussed and bring them back to update my plan. Listing other department goals on my goals sheet forces me to engage them when making my plans. In Figure 6.5, for example, you can see that the Lighting Department gets lightboxes on Monday the 10th. Since The Scenic Department plans to install the lightboxes on Wednesday the 12th, I must ensure that I complete the lightbox project on Tuesday. For this schedule, I planned to utilize a split crew to continue hang in the theatre while siphoning off a team to work on the lightboxes in the electrics shop.

## Load-in & Changeover Plan

Loading In: Pipeline
Loading Out: Winter Rep Plot
Venue: Bingham Theatre

| Date | AM Goals | PM Goals | Labor Needs | External Conflicts/Notes |
|---|---|---|---|---|
| Monday - 11/26 | Plot Review | Schedule Prep/Load-in<br>Draft LED Plan<br>Start Dimmering<br>Submit Rental Quote Requests | JW/TW(in PM) | |
| Tuesday - 11/27 | Order LED Supplies | Finish Dimmering | JW/TW(in PM) | 10a Braden Visits |
| Wednesday - 11/28 | Make Hang Cards<br>Make Hang Limits<br>Make Transportation Sheet | Data and Hot Power Worksheet<br>Mult Sheet<br>Mult Prep<br>Pull Data/Edison<br>Measure Slings<br>Final Rigging Paperwork | JW/TW(in PM) | |
| Thursday - 11/29 | Color Prep | Pull/Label LED Decoders<br>Pull/Label LED Power Supplies<br>Pull Rigging Hardware<br>Pull/Label Pipe<br>Make Stingers | JW/DC/TW(in PM) | |
| Friday - 11/30 | Make Strike Road Case<br>Make Load-in Road Cases<br>Print Paperwork | Hang Stingers on Truss<br>Confirm Rentals | JW/DC/TW(in PM) | |
| Monday - 12/10 | Strike Winter Set-Up | Mult Runs (Don't Drop Added Positions)<br>Hang Grid Fixtures | JW/SB/LK/SY | FIRST DAY ON STAGE!<br>Lightboxes to LX<br>Assemble Score Board On Ground |
| Tuesday - 12/11 | Cable Grid Fixtures<br>Solder and Wire LEDs for Lightboxes | Cable Grid Fixtures<br>Color and Accessorize Grid Fixtures<br>Install LEDs into Lightboxes<br>Lightboxes to Sets EOD | JW/SB/LK/<br>DC/TW(in PM) | Fly Scoreboard |
| Wednesday - 12/12 | Solder Tube and Flying Light LEDs in Shop<br>Install Tube Decoders to Inside of Board<br>Install Power Supplies to Top of Board | Solder Tube and Flying Light LEDs in Shop<br>Install Track on Outside of Board<br>Run Cable/Wire Inside Board | JW/SB/LK/SY/<br>DC/TW(in PM) | 10a MS4 Production Meeting<br>Lightboxes into Scoreboard |
| Thursday - 12/13 | Data and Hot Power Runs<br>Install LEDs to Track | Test and Troubleshoot Scoreboard<br>Transport Lights and Accessories | JW/SB/LK/SY/<br>DC/TW(in PM) | |
| Friday - 12/14 | Rig Added Positions<br>Lower Added Position Mults | Hang, Cable, Color, Accesorize Added Positions<br>Raise Scoreboard | JW/SB/LK/SY/<br>DC/TW(in PM) | 11a PTC Humana Team in BT<br>5p ATLX Meeting |
| Monday - 12/17 | Catch-All | Catch-All | JW/DC/SB | 11a BT Humana Production Meeting |
| Tuesday - 12/18 | Patch<br>Troubleshoot | Console Time | JW/DC/SB/TW (in PM) | |
| Wednesday - 12/19 | Add Rentals<br>Catch-All | Catch-All | JW/DC/SB/WC | |
| Thursday - 12/20 | Catch-All | Catch-All | JW/DC/SB/WC | Set Dressing<br>Pink Noise 6pm |
| Friday - 12/21 | FOCUS | FOCUS | Everyone | |

**Figure 6.5** A typical schedule.

In addition to the goals of the day, I use a "Labor" column to track what staff I will need to complete those goals. For example, if I set a goal for day one of installation to be "Hang Front of House," how many people will it take to do that in one day? Am I planning to do that by myself? Am I planning to use house staff, or will I need to bring in additional help? On this first pass there are two important questions: "How many people do I need?" and "Where are those people coming from?" If you have house staff to use, it is likely that you will use them because they are probably being paid either way. However, they might be assigned to run a show in a different venue at that time and using them might impact that show or require overtime. If you are bringing in additional help, you will need to know how much you will need and how much they will cost.

In this way, we start to think about the Labor Price Out. Labor is intrinsically connected to time, so in a sense we are doing these two price outs simultaneously. Start by indicating your daily goals, then fill out the column of what labor you expect to need to achieve those goals. When you are done, you'll look to see if you were able to fit the project timeline into the time available and if the labor you need matches the labor you have access to. If either does not, look to see if a tweak to one fixes the other. For example, if you planned a week to hang the show with yourself and two lighting technicians, but you only have access to one lighting technician, how long will it take to hang the show? Does that time fit into your timeline? Alternatively, if you planned a week to hang the show with yourself and two lighting technicians but your timeline only allowed three days, how many more lighting technicians do you need to reduce the time to three days? Does that number of technicians fit within your labor budget? If neither of those solutions solve the problem, you will have to look toward materials or record a red flag for a designer conversation about the scope of the design.

Next, we turn toward materials. The Materials Price Out has two components—Design Materials and Infrastructure Materials. The Design Materials are the easiest to plan for. Using the counting features in Lightwright, you can look at what equipment is requested. If the equipment requested is outside the equipment the venue owns, the

### Design Price Out Details

| Show: | Pipeline | | | | Budget: | $ | 7,500.00 |
|---|---|---|---|---|---|---|---|
| Designer: | Alan Edwards | | | | GT Cost: | $ | 2,478.59 |
| Rental Period: | 12/31/18 - 2/6/19 | | | | Budget Less Cost: | $ | 21.41 |

| Project | Item | Qty. | Vendor | Rent/Purch? | Unit Cost | Total Cost |
|---|---|---|---|---|---|---|
| Color | L202 | 2 | BMI Supply | Purchase | $ 7.05 | $ 14.10 |
| Rental | 50 Degree Source 4 Barrels | 11 | Phoenix Lighting | Rental | $ 30.00 | $ 330.00 |
| Scoreboard Boxes - LED Tape | RGB+Daylight White (Spare Tape) | 1 | Environmental Lights | Purchase | $ 137.55 | $ 137.55 |
| Scoreboard LED All Tubes - LED Tape | RGB+Daylight White 4in1 (Includes Spare) | 3 | Environmental Lights | Purchase | $ 189.40 | $ 568.20 |
| Scoreboard LED Long Tubes - Track | Klus 00413L Cover (6.56 ft) | 14 | 1000 Bulbs | Purchase | $ 30.86 | $ 432.04 |
| Scoreboard LED Long Tubes - Track | Klus Anodized Aluminum GIP Channel (6.56 ft) | 14 | SuperBrightLEDs | Purchase | $ 25.95 | $ 363.30 |
| Scoreboard LED Long Tubes - Track | Klus End Cap 00306 | 28 | SuperBrightLEDs | Purchase | $ 1.95 | $ 54.60 |
| Scoreboard LED Long Tubes - Track | Klus 24143 - LED Profile Mounting Clip | 30 | SuperBrightLEDs | Purchase | $ 2.95 | $ 88.50 |
| Scoreboard LED Long Tubes - Track | Discounts | 1 | SuperBrightLEDs | Purchase | $ (66.50) | $ (66.50) |
| Scoreboard LED Short Tubes - Track | Klus 00413L Cover (3.28 ft) | 10 | 1000 Bulbs | Purchase | $ 15.43 | $ 154.30 |
| Scoreboard LED Short Tubes - Track | Klus Anodized Aluminum GIP Channel (3.28 ft) | 10 | SuperBrightLEDs | Purchase | $ 12.95 | $ 129.50 |
| Scoreboard LED Short Tubes - Track | Klus End Cap 00306 | 20 | SuperBrightLEDs | Purchase | $ 1.95 | $ 39.00 |
| Scoreboard LED Short Tubes - Track | Klus 24143 - LED Profile Mounting Clip | 20 | SuperBrightLEDs | Purchase | $ 2.95 | $ 59.00 |
| Templates | Template Reserve | 1 | | Purchase | $ 100.00 | $ 100.00 |
| Wire | Wire Reserve | 1 | | Purchase | $ 75.00 | $ 75.00 |

Figure 6.6 A typical materials price out..

cost of acquiring that equipment is added to your price out. Additionally, it is neccessary to evaluate the need for color filters, templates, as well as any practical or set electric components.

The second component—Infrastructure Materials—can be more difficult to pin down. First, there is the discussion of spare equipment. If a venue has 100 lights but needs 110 for a show, acquiring ten more lights is often not sufficient. A Lighting Supervisor must plan installations that will ensure that every performance from opening to closing looks the same as when the Lighting Designer last saw it. If a light fails, there needs to be a replacement plan built into the infrastructure of the show. This is why it is good to try to limit the amount of equipment made available for the designer during the initial onboarding. Of course, designers will frequently ask for more equipment than you made available and you will have to determine how you will handle that overage. Will you use your spare reserve, will you purchase additional equipment, or will you rent? The decision will ultimately be tied to an evaluation of available budget and your producer's risk tolerance. For most equipment, a 10% spare is sufficient to cover failure potential, but with expensive equipment 10% spare could be several thousand dollars' worth of unused rental equipment.

If a venue has 100 lights and needs 110 for the show, the venue will need to acquire 21 lights to reach 10% spare. Ten lights would be for Design Materials and 11 lights would be for Infrastructure Materials. However, even if the rental cost is only $50 per fixture, you will end up paying $550 of your producer's budget for peace of mind. That might sound worth it, but it also might not be an option. You have to make sure that you understand your producer's risk tolerance. What is more important to them—saving money or ensuring that all the lights will work at every performance?

Further, not all potential issues have the same impact on the production or the budget. For example, the risk of losing one conventional ellipsoidal in a plot of 110 is much less significant than the risk of losing a show-critical moving light whose repair time could easily extend for multiple performances. Therefore, I might be fine with cutting the spare conventional fixtures but would want to hold on to any spare moving lights for much longer. Expendables like lamps and color filters are almost certainly going to fail during a performance run, so the need to mitigate that risk is much higher. Of course, the cost of doing so is extremely low. Carrying 10% or even 50% spare of expendables is affordable for almost all producers.

The second part of Infrastructure Materials is the hidden components of the design. These can include dimmers, cable, tie line, gaff tape, wire, control infrastructure, pipe, and other rigging components. For example, a design could require a light to be in a particular spot and while you may have that fixture in your inventory, you might need to rig a new hanging position to put the light there. There is no cost in using the light since you already own it, but putting the light in that place will add to your Infrastructure Materials budget because you will need to acquire pipe, aircraft cable, and other rigging supplies.

The final part of Infrastructure Materials is connected back to the Time and Labor Price Outs. Your time and labor plan presupposes that you have the equipment you need to do the work properly. In the previous example, in addition to the cost of getting the light to hang in a particular spot, you will need to find a way to get a lighting technician to that spot to do the work. Do you have the appropriate ladder or other access device? Is that ladder in use by another department? Do you have the appropriate safety equipment to get a person to that light? What gear or plan do you have to service that light if it fails during the performance run? If this installation equipment does not exist at your venue, you may have to purchase or rent it. All these elements could make the cost of that one light skyrocket quickly.

Additionally, this final part of the Infrastructure Materials needs to be considered when planning the time and labor for the design. When you determine the timeline and labor needs for your plan, you are doing that with a particular process in mind. Does that process require any equipment that you do not have access to? If it does, the cost of that timeline or labor plan is increased. You may have to consider alternative approaches. Conversely, if your timeline or labor plan is already exceeding available resources, using additional infrastructure materials could solve the problem. For example, if we look at the earlier example where you only had three days to hang a show with two lighting technicians, you may discover that since you only have one ladder that plan is impossible. Purchasing a second or third ladder may allow more work to happen simultaneously and decrease the amount of time the installation takes.

There is a lot of mental processing that goes into the Price Out. It is as if you are performing the installation from start to finish in your head one step at a time. With each step you think to yourself, "what do I need to do that?" or "is there something I can do in advance to make that easier?" If you work this way, you will ultimately come up with your best plan to fit into the parameters the producer has outlined. Inevitably, as you move forward with the rest of the Prep Process, unexpected costs will emerge. The more experience you have, the less this happens, but even the most experienced Lighting Supervisors will be surprised at times. I always plan my schedule conservatively and set aside a small amount of time and money for contingencies. How much you choose to set aside will depend on how risk tolerant you are.

If you get through this process and find that the design does not fit within your resources, you will have to take the design back to the Lighting Designer. Remember, it is not a failure if you cannot do the design as drawn. Many early-career Lighting Supervisors and Master Electricians see the design submission as gospel. They must do it exactly how the Lighting Designer wanted it done or it is a failure on their part. At moments like that it is important to remember that you are collaborating with the Lighting Designer. The work you have done in the Plot Review and the Price Out is part of the process. You have found things that are beyond the Lighting Designer's scope. Likewise, it is important to remember that the Lighting Designer is not trying to slip things past you. If they used too many lights, it is not a conspiracy to bankrupt your producer, they are trying to make the best plot possible.

Take your red flags to the designer. Suggest concrete solutions rooted in the areas that you know are essential to the design from your design meeting conversations. For example, if the design cannot be installed within the time provided, what are the biggest obstacles? Is the designer asking for too many rigging projects? Can more of the house positions be used? If the design needs too many lights that your venue does not have, which ones are most important? Maybe they want color changing ellipsoidals and moving lights. Are they able to take one or the other? Use the listening you did in the meetings to negotiate from a place of understanding. Establish yourself as a partner. Eventually, you will agree on a plan and you can move forward, but make sure that you resolve any Price Out issues before moving forward with prep so that you do not waste time having to repeat parts of the process.

# CHAPTER 7

# Electrical Planning

## CIRCUIT AND DIMMING INFRASTRUCTURE

Once you have a priced-out design that fits within your producer's resources, it is time to start Paperwork Prep. In a sense, you have already begun this process. You have gone through the plot to find any big issues and gotten it ready to be used in the installation. You have prepared a detailed schedule that you can share with other departments. You have ensured that the production will be able to be produced on budget.

However, during Paperwork Prep you will be analyzing the plot in even greater detail. My rule of thumb here is that anything that can be done outside of the installation time should be. This goes double for anything that requires an eye toward safety like electrical or structural planning. I have worked on productions that have not done sufficient Paperwork Prep and the results are almost always negative. Installations can come to a grinding halt while the Lighting Supervisor needs to think through how to rig a new hanging position or cabling needs to be redone because the last circuit and last light are impossibly far away from each other. Always plan these things out, otherwise you can end up backed into a corner with a full crew staring at you, rolling into overtime.

I begin Paperwork Prep with Circuit Planning. Circuit Planning—also called "Circuiting" or "Dimmering"—is a process where you assign dimmers and circuits to each unit in the plot. For Circuit Planning, I only look at units that require a dimmer. Today's lighting designs can have many intelligent fixtures that require data and constant power. I find it much easier to deal with those fixtures separately because those considerations are different than traditionally dimmed fixtures.

I have worked with Lighting Supervisors who prefer to not do Circuit Planning in advance. They will have the installation technicians plug lights into whichever circuits seem convenient in the moment. I strongly recommend against this. Working this way will almost certainly increase your installation time. Whether working in the room or on paper there will be an inevitable struggle to find the last 20% of available circuits. It is much better to have this struggle on paper in advance than to try deal with it in the room with installation time ticking away.

It is important to realize that every venue is structured differently from an electrical standpoint. Before you introduce yourself to the designer, be sure to understand the electrical infrastructure of the venue you will be working in, so that you can properly brief the designer on any considerations that will affect their design. Some Lighting Supervisors will be able to spend years learning the nuances of a single venue whereas others find themselves dealing with new venues on every production. Either way, it is crucial that you take the time to learn how your venue functions as soon as you are brought on to a project. If you discover that the infrastructure of the venue is extremely foreign to you, you will want to engage someone to help you navigate it. This person could be a house staff member if you are a freelancer or another qualified person if you are working with a system that you do not fully understand.

When evaluating your dimming infrastructure, first determine whether it is permanent, temporary, or a hybrid. Most venues that are used for theatre, particularly in the regional theatre, have a permanent dimming infrastructure. This means that dimmer racks are installed in a climate-climate controlled room and connected to individual circuits through conduit and permanent receptacles. Racks, as shown in Figure 7.1, are a tell-tale sign that you are working with a permanent dimming infrastructure.

Outdoor venues and other venues not originally intended for performance may have temporary infrastructure. It is important to note that some temporary infrastructures are not removed regularly and are often treated as permanent. This does not mean that they are permanent for the purposes of this understanding. The crucial difference here is that permanent systems are installed by licensed electricians and have wiring protected by conduit. Temporary systems are often installed by lighting technicians and use temporary cables that can deteriorate with age. In Figure 7.2, you can see "Cam-Lok" connectors in the bottom right of the photograph. These connectors are the tell-tale sign that you are dealing with a temporary infrastructure. You can also see in the photograph that all the dimmed power in the rack is connected to the venue's circuits using "multicable" or "Socapex" rather than wires in conduit. Another clue that this is a temporary infrastructure.

If you inherit a temporary system that is already installed, it is important that you inspect it thoroughly before you first use it and continue to inspect it regularly. If you are not qualified to do so, you should hire a qualified technician to inspect the installation. Similarly, if your production requires you to acquire and install a temporary system as

ELECTRICAL PLANNING  69

**Figure 7.1** A permanent install dimmer rack.

part of your light plot installation, it is important that that work be done by a qualified person. In some jurisdictions, this qualified person must be a licensed electrician. In others, a person with certification like E.T.C.P.'s Portable Power Distribution program or other training may meet the qualifications. Consult your Fire Marshal or Building Inspector for information about your jurisdiction.

It is possible that your system is neither completely permanent nor completely temporary. It could be a hybrid of both. This is different than a temporary system that just is not removed. A hybrid system will have some components that are permanently installed by an electrician. For example, the venue may have circuits that are run in conduit to circuit boxes in the theatre but use portable cable to make the jump from a temporary dimmer rack to an input box. Alternatively, the dimmer racks may be mounted and connected to power through conduit but have circuit boxes in the dimmer room that use portable cable

**Figure 7.2** A portable dimmer rack's telltale Cam-Lok connectors.

to distribute power to the theatre. If you have a hybrid system, it is important to treat the temporary parts of it as a temporary system and inspect them regularly.

Second, you will want to consider the relationship between the dimmers and the circuits. The relationship could be One-to-One, One-to-Multiple, or a Hard Patch. A One-to-One dimmer-circuit relationship is most common in modern theatres. In this relationship, Dimmer 1 is hard-wired to Circuit 1. For theatres with this relationship, the Lighting Supervisor will normally not use circuit numbers in their paperwork because it would result in redundancy—Dimmer 1 will always be Circuit 1.

A One-to-Multiple relationship is a hard-wired relationship where one dimmer is connected to two or more circuits. For example, Dimmer 1 may be connected to Circuit 1 in the back of the house and also Circuit 101 in the back of the stage. This allows the theatre to use Dimmer 1 in two different places without having to run any additional cable. In some cases, these One-to-Multiple relationships can be helpful. In the above example, there may be no lights in the back of the house, so a shared circuit in the back of the stage allows Dimmer 1 to still be used without running a long portable cable across the venue. Of course, this wiring scheme can be equally infuriating. You might find yourself wanting to use Circuit 1 and Circuit 101 on separate dimmers but not be able to because they are hard-wired together. When dealing with One-to-Multiple relationships, the Lighting Supervisor will often include a circuit number in their paperwork in addition to a dimmer number so that the lighting technicians installing the show know whether to plug their cable into Circuit 1 or Circuit 101.

One-to-One and One-to-Multiple relationships are found only in permanent dimming infrastructures. In temporary and hybrid infrastructures, the relationship between the dimmer and circuit is always assignable thought a method known as the Hard Patch. While all temporary systems are Hard Patch systems, this type of relationship is quite common in older permanent systems as well. It was once popular to have circuit counts that far exceeded dimmer counts. Hard Patch relationships allowed for the end-user to distribute the larger number of circuits over the smaller number of dimmers on a show-by-show basis. However, most modern system designers and theatre consultants find it more cost effective to forgo a Hard Patch in favor of more dimmers and a One-to-One relationship. Lighting Designers and Lighting Supervisors alike have appreciated the added flexibility.

The hallmark of the Hard Patch relationship is the patch panel. The patch panel allows the user to determine which circuits are assigned to which dimmers. When using a Hard Patch, the Lighting Supervisor needs to preplan which circuits will be assigned to which dimmers as part of Paperwork Prep. Paperwork that is provided to technicians will include distinct dimmer and circuit numbers. Patch panels come in three varieties—receptacles, pin patch, and slide patch.

As you can see in Figure 7.2, the back of a temporary rack has many receptacles. The rack here has multicable receptacles that can carry six sequential circuits at once as well as single-cable receptacles that can carry individual circuits. Plugging a cable into the Dimmer 1 receptable in the back of the rack assigns that circuit to that dimmer. This is the simplest form of Hard Patch.

The most common type of Hard Patch is the pin patch. As shown in Figure 7.3, the pin patch works like a telephone operator's switchboard. Each circuit is represented by a small patch cable, as seen on the left side of the photo, and each dimmer is represented by a small jack, seen on the right. To patch a circuit to a dimmer, you simply plug the corresponding patch cable into the appropriate dimmer's jack. In a permanent system with a pin patch, there are many hard-wired circuits represented by the patch cables and a much smaller number of dimmers represented on the jack side. This patch system would easily allow you to assign only the circuits that are used in your production to dimmers and easily change the layout for each production.

Most large portable racks, as in Figure 7.2, will use a receptacle system and a pin patch system. In this case, all of output circuits shown on the back of the rack will represented

**Figure 7.3** A pin patch in a portable dimmer rack.

in the pin patch and the end-user will need to assign which ones get which dimmers. For example, the multiconnector outputs we see in the photograph are designed to carry six sequential circuits. In this rack they are numbered A1, A2, and so forth. Here "A" denotes the multicable and "1" denotes which of the sequential circuits is being referenced. Multicable "A" must carry the sequential circuits A1, A2, A3, A4, A5, and A6, but the dimmers assigned to those circuits do not have to be sequential if rack also has a pin patch. In that situation, you can take the patch cable for circuit A1 and plug it to any of the jacks on the dimmer side of the panel. In this way you easily could say that Circuit A1 is Dimmer 3 for one production and then change it to Dimmer 25 for the next production. Circuit A2 could likewise be assigned any dimmer even if it is not sequential with the dimmer used with Circuit A1.

The least common type of Hard Patch is the slide patch panel. This panel is simply a grid made of sliders. On one axis is a list of dimmer numbers. On the other is a list of

**Figure 7.4** A slide patch panel at Armstrong Atlantic State University..
Courtesy of Rob Dillard.

circuits. The sliders move to align a dimmer with a circuit. It is rare to find one of these panels still in use today. They are no longer manufactured for either permanent or temporary systems because modern systems utilize so many circuits and dimmers that the size of one of these grids would be overwhelming. Of course, some older venues may still have one in operation.

Third, you will need to evaluate your system's circuit capacity. Every element of your system will have an electrical rating. These ratings are absolute limits and must never be exceeded. In general, systems are engineered to prevent the end user from exceeding the system's electrical rating, but it is important for the Lighting Supervisor to be aware of these limitations as there are still many opportunities for the end user to create dangerous overcurrent situations particularly when using temporary or hybrid dimming systems.

When evaluating circuit capacity, consider all the elements of the branch. Each branch includes the following elements.

- The dimmer.
- The wiring from the dimmer to the venue—either in conduit or through portable cable.
- The circuit box or other connector in the venue.
- Additional portable cable used between the circuit box and the fixture.

It is important to also remember that each of these elements may have separately rated components. For example, a portable cable is made up of two connectors and the cable itself. It is important that the connectors be rated the same as or higher than the cable.

In the first example branch shown in Figure 7.5 each element has a rated capacity of 20 Amps, thus the branch is rated for 20 Amps. If any of the elements had a capacity less than 20 Amps, the entire branch would need to be rated equal to the lowest capacity of any branch elements. In the second example, all elements have a rated capacity of 20 Amps except the dimmer which is only rated for 15 Amps. In this example the branch would be rated for only 15 Amps.

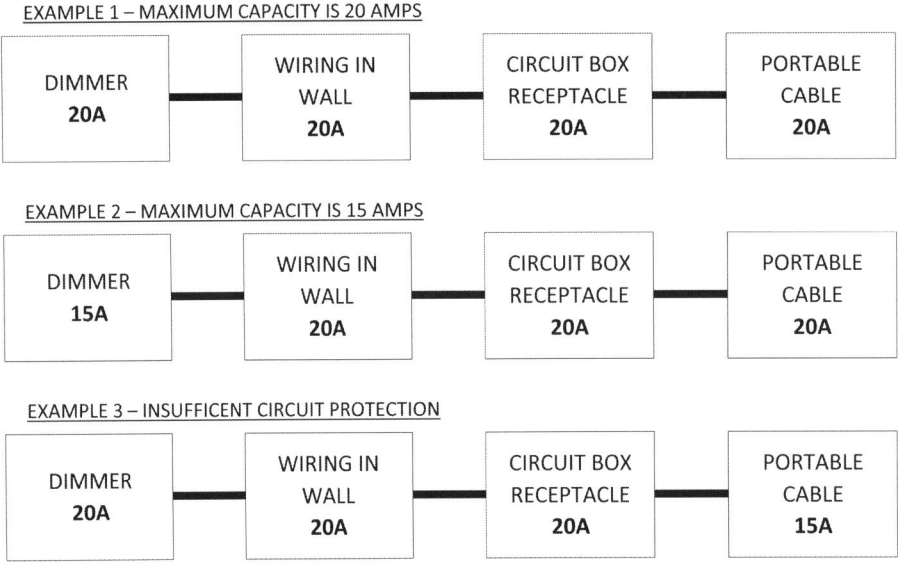

**Figure 7.5** Three branch circuit examples.

In general, the permanent components of your branch will all match. *National Electrical Code* prevents an licensed electrician from running wire in conduit that is only rated for 15 Amps if the dimmer is rated for 20 Amps, for example. However, even in a permanent system, you will still likely need to use some portable cable to get power to your fixtures. This cable will contribute to the composition of the branch. When selecting a portable cable, at a minimum, it must have an ampacity that exceeds the fixture's load. A fixture that draws 10 Amps will need to have a cable that is rated for at least 10 Amps.

The rating of portable cable, or its' "ampacity," is determined by a combination of its' size—expressed as an American Wire Gauge (A.W.G.) number—and its' length. *National Electric Code* sets ampacities for portable cable based on the cable type and size. The first chart in Figure 7.6 lists ampacities for "SO" Cable types. Stage lighting cable is "SOOW." This designation means "Standard" with "Oil Resistant Insulation and Outer Jacket" and "Approved for Outdoor Use." For the most part all portable cable used in theatrical applications should be "SOOW."

Some jurisdictions will allow for the use of "SJOOW"—where the "J" stands for "Junior Hard Service"—in certain applications. Junior Hard Service cable is rated for a maximum of 300 Volts, whereas "SOOW"—also referred to as "Hard Service"—is rated for a maximum of 600 Volts. "SOOW" cable features notably thicker insulation and outer jacket than "SJOOW." This serves to provide superior protection from damage. Since most hardware store extension cords are "SJOOW," it is in abundance at venues. While I generally discourage the use of this sort of cord for your lighting installations, there can be valid uses for it. However, I recommend that you consult your local Fire Marshall before proceeding with any power cable that is not "SOOW." Your local jurisdiction has the right to inspect your work at any point and ask you to remove cable that does not meet their requirements. You do not want to have to cancel a performance while you change out all the cable in your plot.

Length is also a factor for portable cable ampacity. As a cable's length is increased, the resistance of its' copper conductors increase as well. Increased resistance will cause the voltage of the cable to drop and generates heat. With an incandescent load, voltage drop causes the fixture to appear dimmer. With electronic loads such as LEDs or moving lights, voltage drop can cause the electronics to shut down or fail. To protect against these situations, derate the cable's ampacity as its' length increases. The second chart shown in Figure 7.6 provides recommended ampacity limits for "SOOW" cable. These reduced limits ensure that the voltage does not drop by more than 5 volts even if the load reaches 150% of the rated value.

To determine the adequate size cable for yourself, you can use the following formula:

$$(Amps \times Distance\ in\ Feet \times 21.6) \div Desired\ Voltage\ Drop = \text{"circular mils"}$$

"Circular mils" is a measure of the surface area of the cross section of the copper conductors in your cable. Figure 7.7 lists the circular mils for certain cable sizes. As an example of this formula in use, consider a fixture that will draw 10 Amps and is 100 feet away from

## Allowable Ampacity for Flexible Cords and Cables
*(Cable Types SOOW and SJOOW at 85° Fahrenheit)*

| Copper Conductor Size (AWG) | Ampacity of 3-Conductor Cords |
|---|---|
| 18 | 7 |
| 16 | 10 |
| 14 | 15 |
| 12 | 20 |
| 10 | 25 |
| 8 | 35 |
| 6 | 45 |
| 4 | 60 |
| 2 | 80 |

## Recommended Ampacity Limits for Various Lengths and Sizes of SOOW and SJOOW Cable
*(Requires connectors with equal or greater ampacity, voltage drop 5 VDC or less, load 150% of rating)*

| Copper Conductor Size (AWG) | 5 ft | 10 ft | 15 ft | 25 ft | 50 ft | 75 ft | 100 ft | 150 ft | 200 ft |
|---|---|---|---|---|---|---|---|---|---|
| 18 | 7 | 7 | 7 | 7 | 5 | 3 | 2 | 1 | 1 |
| 16 | 10 | 10 | 10 | 10 | 7 | 5 | 3 | 2 | 1 |
| 14 | 15 | 15 | 15 | 15 | 12 | 8 | 6 | 4 | 3 |
| 12 | 20 | 20 | 20 | 20 | 20 | 13 | 10 | 6 | 5 |
| 10 | 25 | 25 | 25 | 25 | 25 | 21 | 16 | 10 | 8 |

**Figure 7.6** Ampacity for SOOW and SJOOW Cords based on size and length.

Ampacity limits are from *National Electric Code* Table 400.4 (A)(1), reproduced with permission of *NFPA* from *NFPA 70®*, *National Electrical Code®*, 2020 edition. Copyright© 2019, National Fire Protection Association. For a full copy of NFPA 70®, please go to www.nfpa.org. NFPA 70®, National Electrical Code®, and NEC® are registered trademarks of the National Fire Protection Association, Quincy, MA.

its' circuit. Since it is good practice to ensure a voltage drop of less than 5 volts, you can plug the numbers in the formula like this:

$$(10 \; Amps \times 100 \; Feet \times 21.6) \div 5 \; Volts = 4{,}320 \; circular \; mils$$

4,320 circular mils exceeds the size of 14 A.W.G. cable—4,107 circular mils. This circuit will need to use 12 A.W.G. cable.

In addition to matching the cable's ampacity with its' load, it is also good standard practice to keep the ampacity of the cable above the rating of the branch circuit's protection,

## Circular Mils for Various Conductor Sizes

| Conductor Size (AWG) | Circular Mils |
|---|---|
| 18 | 1,624 |
| 16 | 2,583 |
| 14 | 4,107 |
| 12 | 6,530 |
| 10 | 10,380 |
| 8 | 16,510 |
| 6 | 26,240 |
| 4 | 41,740 |
| 2 | 133,100 |

**Figure 7.7** Circular mils for various conductor sizes.

Circuit protection is found in the form of breakers and fuses. For most dimmed circuits, protection is at the dimmer level. Looking at the third example in Figure 7.5, a dimmer with a rating of 20 Amps will have built-in circuit protection—usually a breaker—that is rated for 20 Amps. Using a 14 A.W.G. cable with this dimmer—a cable with an ampacity of 15 Amps—can present a hazard since the circuit's protection will not engage if the current exceeds the cable's ampacity. Exceeding a cable's ampacity will cause the conductors to heat up and potentially melt their insulation. If the conductors become exposed, they can spark and start a fire. For this reason, it is best practice to match the ampacity of your cable with the circuit's protection or install additional in-line circuit protection.

To help with this issue, most portable cable used in theatrical applications is 12 A.W.G., which carries a 20 Amp rating. As most dimmers are rated for either 15 or 20 Amps, this cable matches in many applications. Still, it is important that the Lighting Supervisor fully understand the cable they are using in their installation, its' limits, and any potential hazards it can create.

As briefly touched upon, the next important thing to consider is circuit load. Each light used in your plot will have a rating for the current it draws when brought to full by the control console. This is its' load. In many cases the load is expressed in terms of power—Watts—rather than in terms of current—Amps. When evaluating circuit load, you are comparing the load of all the fixtures on a branch with the branch's capacity. As the branch's capacity is expressed in terms of current, we will need to convert the fixture's load from power to current using Ohm's Law, which can be expressed as

$$V = I \times R$$

In this formula, "*V*" stands for "Voltage" or "Energy," "*I*" stands for "Current" or "Amps," and "*R*" stands for "Resistance." This formula is often remembered with the pneumonic, "viral" where the formula is the first three letters. You may notice that this formula does not include power or watts. To convert this law to a formula that can be used for our calculation, you need to know how resistance relates to power. That relationship can be written like this where power is "*P*":

$$R = V^2 \div P$$

Substituting for "R" in our initial formula, you can see:

$$V = I \times (V^2 \div P)$$

Through some rearranging and simplifying, the formula you need emerges:

$$P = I \times V$$

Or more commonly written:

$$P = I \times E$$

Here you can see that power, "*P*" is equal to current times energy, now represented with the letter "*E*." This new formula is easy to remember because it spells the work "pie."

To put the formula in action, consider a fixture with a 750-Watt load in a 120 Volt system.

$$750\ Watts = I \times 120\ Volts$$
$$750\ Watts \div 120\ Volts = I$$
$$I = 6.25\ Amps$$

The fixture in question will draw 6.25 Amps. If that value is less than the branch circuit's maximum capacity and that fixture is the only one on the circuit then it is good to go.

Up to this point everything has been considered an absolute maximum. However, just because a circuit's maximum capacity is 20 Amps, that does not mean it is good practice to load the circuit to that capacity. Loading circuits to their maximum capacity places strain on the system that will eventually result in it to fail. To protect your system, common practice dictates that each circuit load should not exceed 80% of its' maximum capacity. So, if the circuit's capacity is 20 Amps, the circuit load should not exceed 16 Amps. Since the fixture we calculated is only 6.25 Amps, it still fits. In fact, there can be two of these fixtures on this circuit as that would only total 12.5 Amps. However, if you tried to put three fixtures on this circuit you would hit 18.75 Amps. While that is less than the maximum capacity of the circuit, it is not less than 80% capacity, so that plan would be a no-go.

One special circumstance here is multicable. When using multicable, *National Electric Code* requires that you consider Electrical Diversity or Load Diversity. Multicable receives a special code allowance that permits the wiring inside to be smaller than that of a single circuit cable with the same rating. For example, a multicable might be rated for 20 Amps using only 14 A.W.G. wire inside of it.

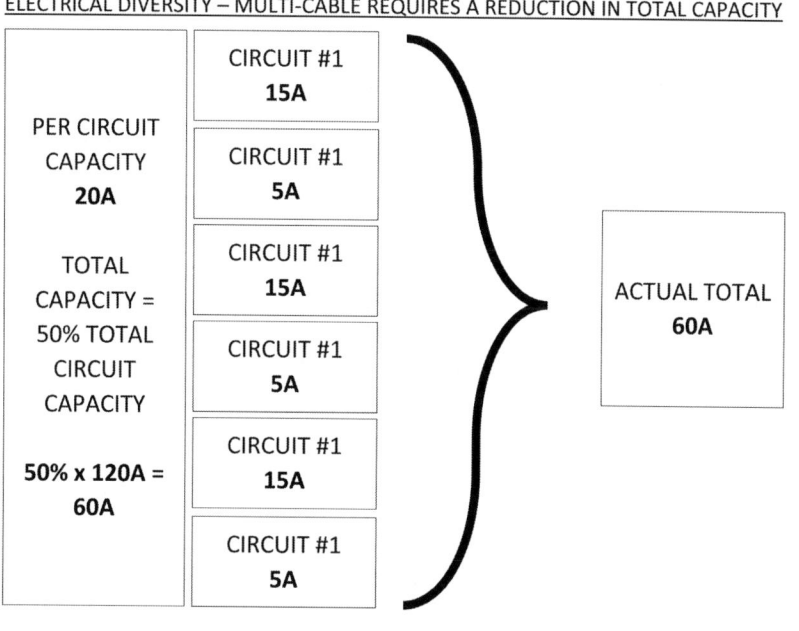

**Figure 7.8** Example of electrical diversity.

This allowance requires the total load over all the branch circuits in the multicable not exceed 50% of the combined circuit ratings. Consider Figure 7.8 as an example. Here, a six-circuit multicable has circuits rated for 20 Amps, the combined circuit rating would be 120 Amps, yet, Electrical Diversity limits that total to 60 Amps. However, the 60 Amps does not need to be distributed equally and each circuit still has an individual capacity of 20 Amps. So, this cable with three circuits at 15 Amps and three at 5 Amps is allowable because the total for the cable does not exceed that 50% diversity nor does it exceed 80% of the individual capacity of each circuit..

The final element to evaluate is phase balance. Most dimming systems in the United States use three phase power. Each dimmer in a rack will be assigned to one of the three phases. In an ideal system, the total circuit load for all the dimmers any one phase of the rack will always be exactly equal to the total load on each of the other phases. If the phases are not in balance, the amount of current on the neutral conductor supplying the dimming rack is exponentially increased which could result in a system failure or the engagement of circuit protection.

Phase balance can be difficult to evaluate because lighting levels change during a performance. This shift in level causes the circuit loads to change. For example, the 750-Watt load above is only equal to 6.25 Amps when it is at full on the control console. If the same light is at 25% on the control console, the current is much less. Consequently, using the maximum capacity of each circuit alone to evaluate phase balance is not sufficient to ensure a safe system.

One of the most common methods of combatting phase imbalance is that each dimmer rack is normally supplied with a neutral conductor with a greater ampacity than the phase hot conductors. This allows built-in tolerance for phase imbalance. Permanent installations would be built with this consideration in mind. If you are working with a temporary or hybrid system, you should ensure this is the case as part of your inspection.

While the upsizing of this conductor will mitigate some imbalance, care still must be taken when assigning loads to phases. In addition to adding up the total maximum circuit load on each phase, consider the purposes of those loads. For example, if Phase A is all front light, Phase B is all back light, and Phase C is all cyc light, it is likely that the phases will always be out of balance because in mixing their looks the designer will want to have each of those systems at different levels. However, if you divide each of those systems so that each phase had one-third of the systems, the designer's level mixing would have less of an impact on the phase balance.

Phase balancing is one of the biggest advantages of a Hard Patch system because you can assign phases to lights on a show-by-show basis and increase your ability to have a balanced system. If you have a system where the dimmers are hard-wired to their circuits, extra care must be taken when assigning circuits as always choosing the closest circuit to a fixture may result in out-of-balance phases. This is where knowing your venue's system infrastructure comes in handy. Most, if not all, permanent systems will be intentionally designed so that the phases of the racks are spread evenly around the building. Instead of all your front of house circuits being hard-wired to Phase A, you would have every third circuit wired to Phase A, then its' neighbor wired to Phase B, and so on. If you look closely at the rack shown in Figure 7.1, you can see how the dimmers are numbered to correspond with their phases instead of sequentially.

## DIMMERING THE PLOT

Once you have a firm understanding of the infrastructure available to you, you begin the process of fitting the plot into the infrastructure. Using software like Lightwright and Vectorworks, you can ensure your work is accurate, comprehensive, and able to be clearly expressed to the lighting technicians who will install the show.

Just as you saw in the plot clean-up process, the efficiency and effectiveness of your work is much improved by your ability to use template Lightwright and Vectorworks Files set-up for your venue. To aid in circuiting, the template Vectorworks File should include a circuiting layer that shows how any permanent infrastructure is laid out and any information you might want quick at hand about it such as the maximum capacity of a branch circuit.

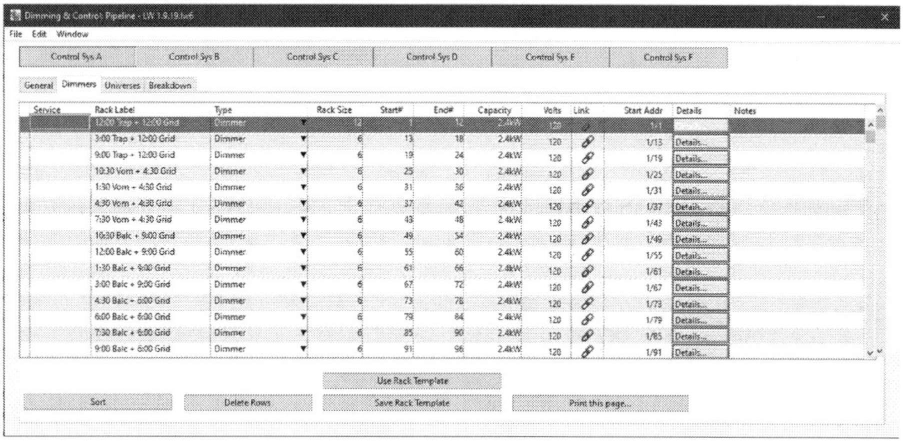

**Figure 7.9** Lightwright's "Dimming & Control" dialogue.

Similarly, Lightwright has a set-up screen called "Dimming & Control." This set-up will allow you to describe your dimming system and its' capacities to the program, so that it can help you calculate your loads. Remember, when entering capacities in this area to reduce them to 80%. This allows for the headroom we discussed in the previous section to be automatically considered. The "Dimming & Control" dialogue also gives you an ability to assign phases to your dimmers and produce a report about your estimated phase balance.

Finally, I prepare a document called a Patch Sheet to help me track dimmer and circuit usage while I am dimmering. This is a scratch sheet that provides a quick visual reference of the circuit and dimming infrastructure. The example venue described in Figure 7.10 has six hanging positions. Each has 18 circuits. You can see that each position's circuits are represented by a box. When I assign a circuit to a fixture, I will place an "X" through the box in pencil so that I know that circuit has been used.

If the venue used a one-to-one patch relationship, I would only need this top section on my Patch Sheet because each circuit would automatically correspond to a particular dimmer. However, the venue represented by Figure 7.10 has only 96 dimmers and requires a Hard Patch to assign each circuit to a dimmer, so I have added a section to the bottom to represent the dimmers in the rack.

Once I have completed the top section and assigned each light a circuit, I can then use the bottom section to assign each circuit a dimmer. Just as I did with circuits, I can place an "X" through each dimmer box with my pencil so that I know that dimmer is used. You will also notice that the dimmer section is shaded to show which dimmers are on which phase. This can help illustrate any phase imbalance. For example, If I only used 36 dimmers in my planning, I can see that I would not want to just use dimmers 1 through 36 because I would be heavily loading Phase A and not putting any fixtures on Phases B or C.

## PATCH SHEET

FOH 3

| A 1 | A 2 | A 3 | A 4 | A 5 | A 6 | A 7 | A 8 | A 9 | A 10 | A 11 | A 12 | A 13 | A 14 | A 15 | A 16 | A 17 | A 18 |

FOH 2

| B 1 | B 2 | B 3 | B 4 | B 5 | B 6 | B 7 | B 8 | B 9 | B 10 | B 11 | B 12 | B 13 | B 14 | B 15 | B 16 | B 17 | B 18 |

FOH 1

| C 1 | C 2 | C 3 | C 4 | C 5 | C 6 | C 7 | C 8 | C 9 | C 10 | C 11 | C 12 | C 13 | C 14 | C 15 | C 16 | C 17 | C 18 |

ELECTRIC 1

| D 1 | D 2 | D 3 | D 4 | D 5 | D 6 | D 7 | D 8 | D 9 | D 10 | D 11 | D 12 | D 13 | D 14 | D 15 | D 16 | D 17 | D 18 |

ELECTRIC 2

| E 1 | E 2 | E 3 | E 4 | E 5 | E 6 | E 7 | E 8 | E 9 | E 10 | E 11 | E 12 | E 13 | E 14 | E 15 | E 16 | E 17 | E 18 |

ELECTRIC 3

| F 1 | F 2 | F 3 | F 4 | F 5 | F 6 | F 7 | F 8 | F 9 | F 10 | F 11 | F 12 | F 13 | F 14 | F 15 | F 16 | F 17 | F 18 |

| 41 | 42 | 43 | 44 | 45 | 46 | 47 | 48 | 89 | 90 | 91 | 92 | 93 | 94 | 95 | 96 |
|----|----|----|----|----|----|----|----|----|----|----|----|----|----|----|----|
| 33 | 34 | 35 | 36 | 37 | 38 | 39 | 40 | 81 | 82 | 83 | 84 | 85 | 86 | 87 | 88 |
| 25 | 26 | 27 | 28 | 29 | 30 | 31 | 32 | 73 | 74 | 75 | 76 | 77 | 78 | 79 | 80 |
| 17 | 18 | 19 | 20 | 21 | 22 | 23 | 24 | 65 | 66 | 67 | 68 | 69 | 70 | 71 | 72 |
| 9  | 10 | 11 | 12 | 13 | 14 | 15 | 16 | 57 | 58 | 59 | 60 | 61 | 62 | 63 | 64 |
| 1  | 2  | 3  | 4  | 5  | 6  | 7  | 8  | 49 | 50 | 51 | 52 | 53 | 54 | 55 | 56 |

**Figure 7.10** Example Patch Sheet.

With the files and scratch sheet ready, you can set to work. Begin by assigning circuits to fixtures. Consider each circuit's capacity and its' load when assigning these circuits. If you have a Hard Patch system, hold off on dealing with dimmers until you have assigned all the circuits. As you assign a circuit, enter the information in the appropriate Lightwright column and cross off that circuit on your patch sheet. The goal when selecting a circuit

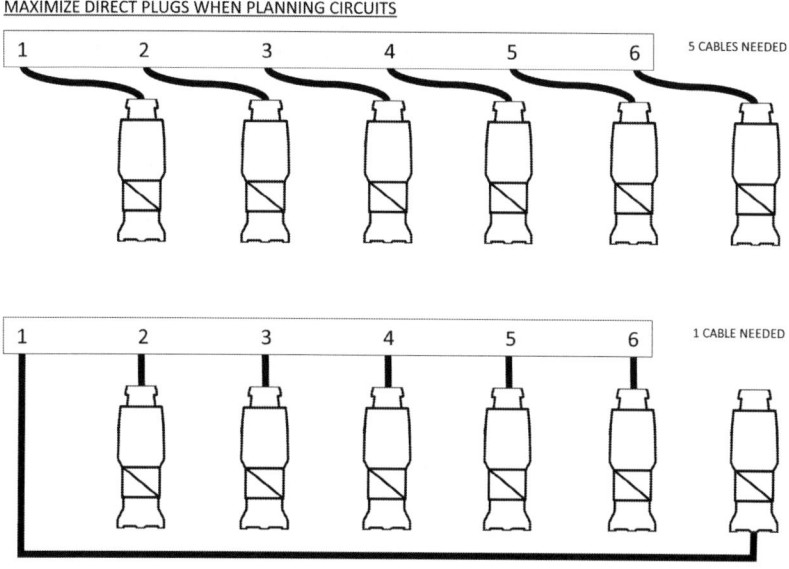

**Figure 7.11** Each fixture in the first example is too far from its' circuit to be a direct plug, so each will need a short cable. In the second example, every fixture but one is able to be a direct plug and only one fixture needs a cable.

is to require the least amount of portable cable be installed in the field. When I plan my circuits, I look for "direct plugs" first. These are fixtures that can plug directly into a circuit without any portable cable. We will discuss this more in the next chapter when we talk about "focus slack" and cable dressing, but for now remember that the furthest a fixture can be from its' circuit and still be a direct plug is about 18 inches.

In Figure 7.11, for example, there are five lights and a connector strip. The circuits and the lights are both spaced at 24 inch intervals, but the two rows are also offset by 24 inches. It may seem logical at first to assign the first light to the first circuit and so on down the line. However, since that will put each light 24 inches from its' circuit, none of them will be able to be direct plugs. Each fixture will require a short portable cable to reach its' circuit. Instead, if we make four direct plugs and leave one light 120 inches from a circuit, we only require one cable to be run. Even though this cable is relatively long, it will still be faster to install it than five shorter cables. Additionally, limiting this plan to only one portable cable reduces the number of potential failure points in the installation as the more equipment you use the more failure points you have. A good rule of thumb here is that running one 50-foot cable is always better than 50 one-foot cables.

If you are using multicable—either from a portable rack in a temporary installation or to carry a group of permanent circuits elsewhere in the venue—it is important to make sure your plans for this cable is clearly documented as well. To track my use of multicable, I make a "Mult Sheet." This sheet—like the one shown in Figure 7.12—describes where

## Mult Prep

| Show: | BT Humana 2019 | | | | | | | | |
|---|---|---|---|---|---|---|---|---|---|
| Venue: | Bingham Theatre | | | | Yellow Highlight = Needs Prepped | | | | |
| Designer: | Heather Gilbert | | | | | | | | |

| Mult | BI | BO | Circuits | Spare | Hot Power | Origin | Drops to | Length |
|---|---|---|---|---|---|---|---|---|
| A | Soco | 8' Staggered | 209-214 | | | 12:00 Annex in BUG | 10:30 Truss (9:00 Side) | 60+20 |
| B | Soco | 8' Staggered | 215-220 | | | 12:00 Annex in BUG | 10:30 Truss (12:00 Side) | 60+20 |
| C | Soco | 8' Staggered | 221-226 | | | 12:00 Annex in BUG | 1:30 Truss (12:00 Side) | 50 |
| D | Soco | 8' Staggered | 227-232 | | | 12:00 Annex in BUG | 1:30 Truss (3:00 Side) | 50 |
| E | Soco | 8' Staggered | 233-238 | | | 12:00 Annex in BUG | 4:30 Truss (3:00 Side) | 75 |
| F | 2' BI | 8' Staggered | 1-6 | | | 12:00 BUG | 4:30 Truss (6:00 Side) | 75 |
| G | 2' BI | 8' Staggered | 21, 8-12 | | | 12:00 BUG | 7:30 Truss (6:00 Side) | 75 |
| H | 2' BI | 8' Staggered | 41-42, 45-48 | | | 4:30 BUG | 3:00 Bridge 2, #8 | 40 |
| I | 2' BI | 8' Staggered | 90-93, 95-96 | | | 6:00 BUG | 6:00 Bridge 4, #5 | 30 |
| J | 2' BI | 8' Staggered | 66-68, 70-72 | 67-68 | | 9:00 BUG | 9:00 Bridge 5, #6 | 30 |
| 5K | N/A | N/A | 195 | | | 5K Box Above Grid Door | 1:30 Bridge 1A, #4 (via BUG) | 50 |
| 5K | N/A | N/A | 196 | | | 5K Box Above Grid Door | 4:30 Bridge 1A, #10 (via BUG) | 75 |
| 5K | N/A | N/A | 197 | | | 5K Box Above Grid Door | 9:00 Bridge 3, #5 (via BUG) | 75 |
| 5K | N/A | N/A | 198 | | | 5K Box Above Grid Door | 1:30 Truss, 12:00 Side (via BUG) | 50+15 (on truss) |

**Figure 7.12** A typical Mult Sheet.

each multicable run will originate, where it will run to, how long it will be, and which circuits or dimmers it will carry. For example, in Figure 7.12 you can see that "Mult A" will begin at the "12:00 Annex in BUG" and run to "10:30 Truss (9:00 Side)." It will carry circuits 209 through 214. I can use this paperwork as a quick reference of the multicable used in the production for further price outs, installation, and troubleshooting.

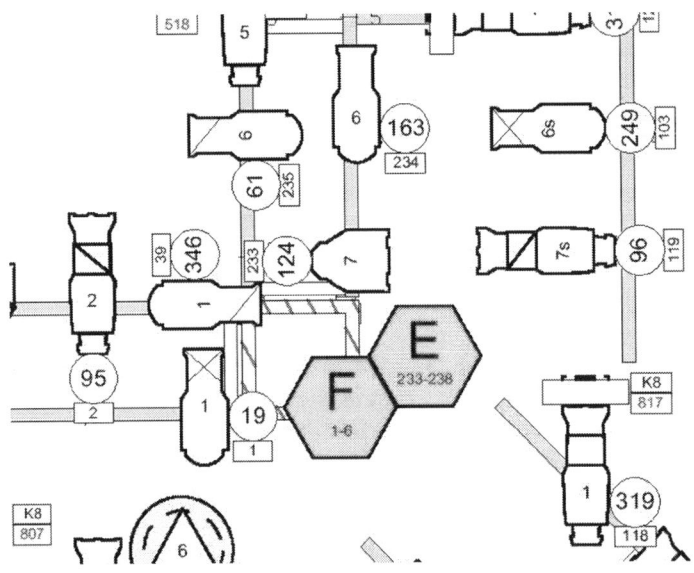

**Figure 7.13** Multi-cable termination locations drawn in Vectorworks.

| Position | Unit# | Instrument Type | Wattage | Purpose | Color | Gobo | Channel | Address | Dim/HP | Mult | Leg# | Cabling |
|---|---|---|---|---|---|---|---|---|---|---|---|---|
| Truss 5 | 1z | ETC S4-26 | 575w | DMLC | R132 | | (86) | 1/217 | 217 | B | 5 | |
| Truss 5 | 2 | ETC S4-PAR NSP | 575w | CW < Tight | N/C | | (210) | 1/15 | 15 | | | BUG @ 12 (Break at Truss Corner) |
| Truss 5 | 3 | Altman 6x9 | 750w | Frnt Cool P | L200+R119 | | (154) | 1/218 | 218 | B | 4 | |
| Truss 5 | 4 | ETC S4-PAR MFL | 575w | Blue Down MDC | L200 | | [illegible] | [illegible] | [illegible] | A | 1 | [illegible] |
| Truss 5 | 5 | ETC S4 PAR NSP | 575w | CW: Tight Down | N/C | | (236) | 1/223 | 223 | C | 3 | |
| Truss 5 | 6 | ETC S4-26 | 575w | DMRC | R132 | | (87) | 1/222 | 222 | C | 2 | |
| Truss 5 | 7 | Altman 6x16 | 750w | Vom: 7:30 | L202+R119 | | (17) | 1/221 | 221 | C | 1 | |
| Truss 6 | 1 | Altman 6x12 | 750w | HTD 10:30 Vom | R114+L202 | | (348) | 1/20 | 20 | | | BUG @ 12 |
| Truss 6 | 2 | ETC S4-19 | 575w | CW: Chair | L201+R119 | | (248) | 1/210 | 210 | A | 2 | |
| Truss 6 | 4 | ETC S4-36 | 575w | Down DC | R132 | | (102) | 1/219 | 219 | B | 5 | |
| Truss 6 | 7 | Altman 6x12 | 750w | HTD: 1:30 Vom | R114+L202 | | (345) | 1/19 | 19 | | | BUG @ 12 |
| Truss A | 1 | ETC S4-PAR MFL | 575w | Blue Down MDR | L200 | | (122) | 1/209 | 209 | A | 1 | 3Fer w: T5#4/TF#7 |
| Truss A | 2 | Altman 6x9 | 750w | Frnt Cool Q | L200+R119 | | (156) | 1/228 | 228 | D | 2 | |
| Truss A | 4 | ETC S4-26 | 575w | MR | R132 | | (92) | 1/229 | 229 | D | 3 | |
| Truss A | 6 | Altman 6x9 | 750w | Bx Cool C | L200+R119 | | (163) | 1/234 | 234 | E | 2 | |
| Truss A | 7 | ETC S4-PAR MFL | 575w | Blue Down MUR | L200 | | (124) | 1/233 | 233 | E | 1 | 3Fer w/ T2#4/TF#1 |
| Truss B | 1 | ETC S4-PAR NSP | 575w | CW: Prayer Special | N/C | | (246) | 1/227 | 227 | D | 1 | from 1:30 Mult |
| Truss B | 2 | Altman 6x12 | 750w | N | L201+R119 | | (33) | 1/224 | 224 | C | 4 | |
| Truss B | 3 | ETC S4-10 | 575w | PTC: Speaker Special | R132 | | (402) | 1/225 | 225 | C | 5 | |
| Truss B | 5 | ETC S4-19 | 575w | PTC: Squares | L200 | | (406) | 1/236 | 236 | E | 4 | |
| Truss B | 6 | Altman 6x12 | 750w | A | L201+R119 | | (61) | 1/235 | 235 | E | 3 | |
| Truss C | 3 | ETC S4-10 | 575w | PTC: Tent Down | R119 | | (403) | 1/230 | 230 | D | 4 | |
| Truss C | 5 | ETC S4-PAR MFL | 575w | Blue Down C | L200 | | (123) | 1/238 | 238 | E | 6 | |
| Truss C | 6z | ETC S4-19 | 575w | PTC: Squares | L200 | | (408) | 1/3 | 3 | F | 3 | |
| Truss CD 6:00 | 4 | Altman 6x9 | 750w | CW: Bench | L202+R119 | | (227) | 1/237 | 237 | E | 5 | |
| Truss D | 3 | ETC S4-36 | 575w | CW: Center Down | R132 | | (247) | 1/212 | 212 | A | 4 | |
| Truss D | 5 | Altman 6x12 | 750w | CW: Table Down | R132 | | (245) | 1/231 | 231 | D | 5 | from 1:30 Mult |
| Truss D | 6z | Altman 6x9 | 750w | CW: Bench | L202+R119 | | (224) | 1/220 | 220 | B | 6 | |

**Figure 7.14** Dimmering in Lightwright.

In addition to keeping the "Mult Sheet," I find it helpful to record the termination point of each multicable on the hang plot. This helps provide a visual reference as to the location of each of these temporary runs. Since my template drawing also displays the location of each permanent circuit, I try to make sure that the designation for multicable termination is distinct to avoid confusion later.

It is also important to record the multicable information into Lightwright, so that it can be easily accessible during the installation. For this I use separate columns for multi-cable names and legs as shown in Figure 7.14. These are non-default columns that I have configured in my template file using Lightwright's column editor. Editing existing columns or using the "User" columns is a great way to add additional data your paperwork. For example, you can see in Figure 7.14 that I use a "Cabling" column to convey notes to the installation technicians. I have also used user columns to track which equipment is house equipment and which is a rental. The more information you tell Lightwright the more powerful it can be.

Finally, once all the circuits are assigned, you can shift to dimmers. If you are working with a One-to-one system, you are already done. Since each circuit is hard-wired to each dimmer, the circuit number and the dimmer number are the same. In these venues, you can just copy Lightwright's circuit number column into the dimmer column. When I know that all of my circuit numbers and dimmer numbers will be identical, I often repurpose the circuit number column for other uses like multicable and enter my circuit information directly into the dimmer column when planning which is what you see in Figure 7.14.

If you have a Hard Patch, however, the circuits will not be inherently connected to any dimmers and you must make those selections. Assign each circuit to a dimmer paying attention to the total load of each phase as well as the purpose of the loads on each phase. Make every attempt to keep the purposes diverse and the total loads equal. Record the dimmer number in Lightwright and cross it off on your patch sheet.

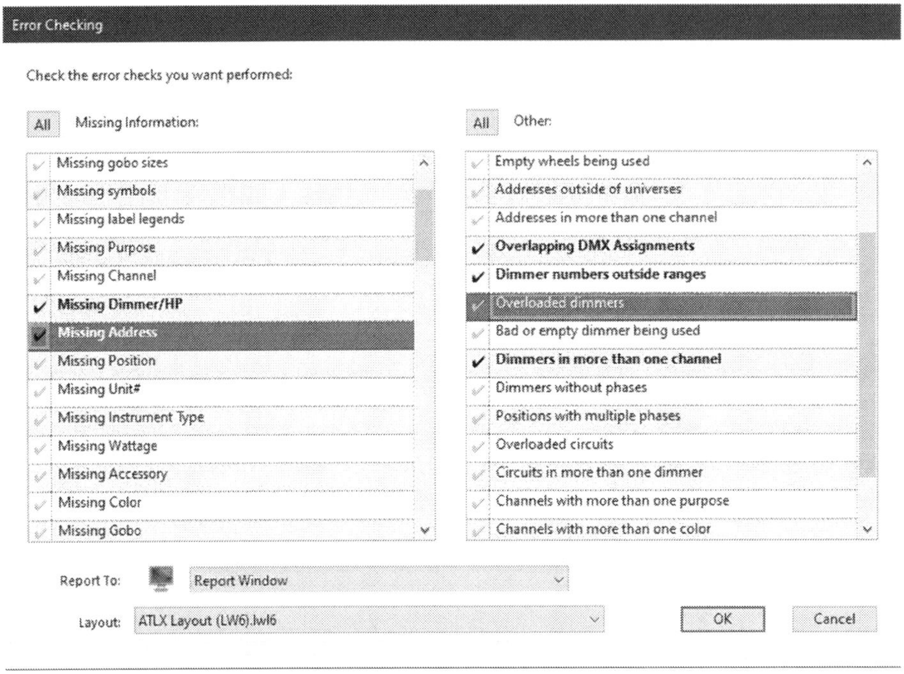

**Figure 7.15** Lightwright's "Error Checking".

In addition to phase balance, also pay attention to the Lighting Designer's channel numbers. A single channel number can control any number of dimmers. However, a dimmer can have only one channel assigned to it. For example, if you had two lights on two circuits with the same channel number, you could give them the same circuit or dimmer provided they fit within that branch's capacity. However, if you had two lights with different channel numbers they must have different circuits and dimmers even if they are well under their branches' capacities.

Finally, when all the dimmers and circuits are recorded in Lightwright, use its' built-in error checker and report engine to look for overloaded dimmers, multiple dimmers in the same channel, multiple dimmers on the same circuit, missing information, and out of balance phases. Really explore these error checking features of the software as they can save you tons of time.

## HOT POWER AND DATA INFRASTRUCTURE

Once Circuit Planning is complete, you can turn your attention to non-dimmed loads. These loads include moving lights, LED fixtures, atmospheric machines, and other devices. The two big identifiers of these loads are that they require unregulated constant power and that they usually receive a direct control signal. For example, the dimmer racks themselves are a non-dimmed load. They receive constant unregulated power—either hard-wired or via a tie-in—and they receive direct control from the console. Like Circuit Planning, Hot Power and Data Run planning must start with a thorough understanding of the existing infrastructure of the venue.

First, we look at Hot Power. "Hot Power" is the common phrase used to refer to unregulated constant power. The outlets in your home, for example, are hot power. The key distinction between dimmed power and hot power is that hot power is unregulated. "Regulation" means that the sine wave of the alternating current is trimmed to maintain a constant voltage range. Dimmed power—even when set to full, "non-dim," or "switched"—retains this trim. Since many electronic devices require an unaltered sine wave to operate properly, using a dimmer to power these devices can cause unexpected problems and potentially damage the equipment. Always power your electronic equipment with unregulated constant power.

As part of your venue evaluation, you need to determine if it is wired for Dedicated Hot Power or Shared Hot Power. Most modern venues have permanently installed receptacles—as shown in Figure 7.16—that are dedicated for lighting power. This is the gold standard because it is easy to know what the current load of any circuit is if you are the only one using it.

**Figure 7.16** Dedicated hot power.

If the venue is older, it is likely that dedicated lighting hot power was not part of the installation. Luckily, there are numerous options to handle this situation. The most obvious one is to use shared building power—the outlets in the wall. In general, there is nothing electrically wrong with this process; however, an abundance of caution must be used here because lighting equipment typically has high current draws. These high currents can quickly reach the maximum circuit capacity of any of these shared circuits especially if they have more than just lighting equipment on them.

When evaluating these circuits, be sure to know where all the receptacles on a particular circuit branch are located. Just because there are two wall outlets, for example, does not mean that you have two circuits. If you are dealing with shared power, it is more likely that multiple outlets share the same circuit than not. Second, it important to know where the breaker panel for each shared circuit is located before you use them. This will allow you to reset the breaker should it be tripped as well as allow you to check the rating for each breaker. Finally, be sure to consider any nonlighting loads that may be on the circuit any time during production. Perhaps you are using an outlet to power a moving light that is often used by Stage Management to vacuum. You will need to consider the total load of the moving light and the vacuum to use that circuit, or you will have to be sure to ask Stage Management to find a different power source.

A second option is using the dimmed circuit infrastructure to send hot power. Because the dimmers in that system are regulating the voltage, you will have to bypass the dimmers in question to use the circuit infrastructure. Fortunately, the dimmer manufacturers will often provide a means for doing this. For Hard Patch systems, instead of patching a circuit to a dimmer, the patch panel or rear of the rack may have section labeled "Hot Pockets" or "Convenience Outlets" like the one shown in Figure 7.17. These are unregulated circuits in the rack that can provide hot power to the existing branches in the system instead of dimmed power. For example, you may have a multicable from a portable dimmer rack with circuits A1, A2, A3, A4, A5, and A6. On the pin patch panel, you can patch A1, A2, A3, A4, and A5 to dimmers 1 through 5 respectively but then patch A6 to Hot Pocket 1. Now leg A6 of your multicable has hot power.

As another bypass option, most dimmer racks—both permanent and temporary—have removable dimmer modules. The manufacturer will likely make optional replacement modules for a variety of applications. Electronic Theatre Controls' Sensor Dimming system, for example, has replacement modules that can act as constant power or relay-controlled constant power. Changing the modules is a relatively simple process but must be done while the rack is fully powered off and with the power locked out. Changing a dimmer module with the rack powered on can result in an arc flash—an electrical explosion caused by current traveling in open air. Arc flashes can result in serious personal injury or death in addition to significant equipment damage. If you need to change the modules in your dimmer rack, be sure to discuss powering down the rack with a qualified individual prior to attempting.

# ELECTRICAL PLANNING 89

**Figure 7.17** Convenience outlets in a portable dimmer rack.

Regardless of what kind of hot power you have, you will need to evaluate the circuit capacity and load requirements in much the same way you did for dimmed power. However, the process for hot power can be slightly more complex as moving lights, LED units, or other electronic devices have more considerations than traditional incandescent lighting fixtures when determining their load.

The VL1100 from Phillips Vari*Lite, for example, takes a 1,000-Watt lamp. Even though the primary light source is incandescent, it would be a mistake to assume the fixture only requires 1,000 Watts or about 8 Amps. A quick review of the fixture's power requirements in its' manual reveals that it actually requires 10 Amps at 115 Volts. This extra current is due to the fixture's motors which need power too. Always consult a fixture's manual to determine its' maximum current draw and other power requirements. If the specifications seem unclear or confusing it is good practice to consult the manufacturer directly.

As for Data Runs the same infrastructure questions apply. You will likely either have a built-in data distribution system or you will be installing a temporary one for the production you are working on. If using an existing one, you need to evaluate how it is set-up and what its' limitations are. Most systems use the DMX-512 standard, which allows 512 control channels to be sent up to 1,000 feet on DMX-data-grade cable. Along these 1,000 foot runs, up to 32 devices can be "daisy-chained" together with cables going into an "input" port, out an "output" port, and then on to the next device (Bennette 2008).

While simple systems might just have one DMX cable connecting a control console and a dimmer rack, most modern systems utilize a complex data infrastructure.

Your data infrastructure—whether permanent or temporary—will either use a fully DMX system or a combination of DMX and network protocols like sACN or ArtNet. As most lighting equipment ultimately needs a DMX signal, it is not currently possible to utilize network protocols exclusively. At some point, the signal will need to be converted to the DMX protocol prior to connection to the lighting equipment. It is important to determine what type of signal your console originates, how it is distributed, and where it is translated to the DMX protocol.

When planning or analyzing a data infrastructure, you should first determine how your control will originate and how much control will be needed. Most of the time, your control originates with a lighting console. The lighting console will output DMX, sACN, Art-Net, or another protocol. The amount of control you will need will depend on the light plot. For example, a show with only conventional fixtures and dimmers will need only channel of control per dimmer. If you have less than 512 dimmers and your console can output DMX, you can simply run a DMX-data-grade cable to the first dimmer rack and then use additional cables to "daisy-chain" other racks together.

However, if your control needs exceed 512 channels, you will have to determine how many DMX outputs your console has the capacity to handle and how many your production needs. Each console type will have a maximum output capacity listed in terms of "addresses." Each unit in the light plot will have a "DMX Footprint" or count of addresses that it needs to be controlled. A dimmer, for example, has a DMX Footprint of 1 because it only needs to be turned on and off. A moving light, on the other hand, can easily have a DMX Footprint of 30 or even 100 as each parameter can require one or more addresses. When evaluating your plot, you will need to determine whether the total size of your DMX Footprint fits withing your console's capacity. For example, Electronic Theatre Control's Ion Xe can have up to 12,288 outputs. If you have a light plot with 100 Robe Spiiders—a fixture with a DMX Footprint of 123—your console will not be able to control all of them because their total DMX Footprint will be 12,300.

Still, even if your count of Robe Spiiders was only 99, leaving you with a DMX Footprint of 12,177, you may not be able to control all your units with the infrastructure that you have. Each console will have a certain number of DMX Output Ports. Each port can send only one universe—512 consecutive control channels—worth of data. The Ion Xe in this example has only four DMX Output Ports. This means it can only send 2,048 channels worth of DMX signal without any additional networking. As a further challenge, each unit in the plot cannot receive more than one universe. So, as an example, if Output Port Number 1 sends Universe Number 1 to the first Spiider, that unit will receive "address" 1 through 123. That Spiider can daisy-chain to the next Spiider which receives address 124 through 246. Then on to the next Spiider which would receive address 247 through 369 and finally on to Spiider Number 4 which receives address 370 through 492.

Since there are only 20 addresses left in the 512 addresses you started with, you cannot add a fifth Spiider to your daisy-chain. Spiider Number 5 will have to be the first unit that is connected to DMX Output Port Number 2 which can be set to output Universe Number 2 starting with address 513. The remaining 20 addresses in Universe Number 1 are simply left unused and do not count against your total output count.

Now, if the structure of the console limits you to only four Spiiders per output port and you only have four output ports, then what is the point of having a console with a capacity of 12,288 outputs? This is where network protocols come in. In addition to consoles, your data infrastructure can include intermediate control devices called "gateways." These gateways connect to your console via a TCP/IP network. Each gateway receives data for all possible output channels that the console has. Then, the gateway is set to output a certain universe of data on one of its' output ports. This means that we can increase the amount of available output ports our console has by using the console as a data hub and the gateways as the distribution method.

Looking back at our 99 Spiiders, we determined that four units fit on one DMX universe. To control all 99 units, you will need to have 25 DMX Output Ports each with a distinct universe. If you are lucky to have seven or more DMX gateways like the rack seen in Figure 7.18, you can then easily control all 99 Spiiders. In this infrastructure, your Ion Xe will output a network protocol. For all the products in Electronic Theatre Control's

**Figure 7.18** An sACN network hub.

family of consoles this protocol is called "ETCNet" which is a variation of sACN. That protocol will be sent to a network switch that duplicates the signal and sends it to each of the seven four-port gateways on your network. You will assign each of the 28 DMX Output Ports a unique universe which gives them each a unique range of DMX addresses. You will then daisy-chain four Spiiders off of each port allowing you control over all 99 units.

Even if you are not trying to control 99 Robe Spiiders, you will still find benefit in some level of data infrastructure. As a matter of preference, I use multiple DMX universes intentionally when they are available, so that each type of equipment can have its' own cable run and own universe. While the DMX-512 standard allows for up to 32 devices to be daisy-chained on a single run, each device represents an opportunity for failure so reducing the number of devices on each daisy-chain can limit potential issues.

Looking at Figure 7.19 for example, you can see a console with four outputs like the Ion Xe discussed above. Here I am assigning Output Ports 1 through 4 to Universes 1 through 4 and then using Universe 1 for dimmers, Universe 2 for moving lights, Universe 3 for LEDs, and Universe 4 for atmospherics like fog machines.

We can even maintain this isolation when the console has only one output. In that case, we can use an additional device called an "opto-splitter." This is a device will take one DMX Output Port and reproduces it over multiple ports. Unlike the gateway which receives all possible outputs from the console, an opto-splitter uses the DMX Protocol so it can only receive the 512 channels from whichever universe is plugged into it. Figure 7.20 uses an opto-splitter with a single port console. Here each unit type still has a dedicated DMX run, but all four-unit types are on the same universe.

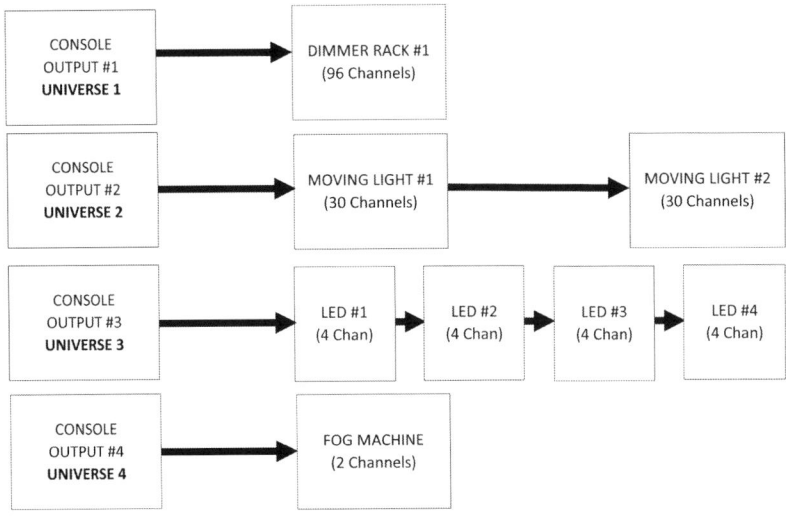

**Figure 7.19** Each type of unit has its' own run back to the console and its' own universe.

ELECTRICAL PLANNING 93

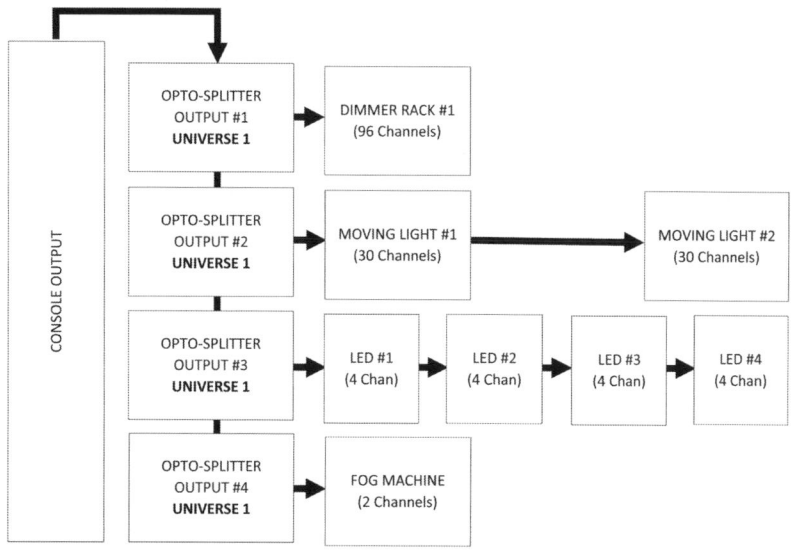

**Figure 7.20** An Opto-Splitter clones the signal from one console output and sends the same universe over multiple cables.

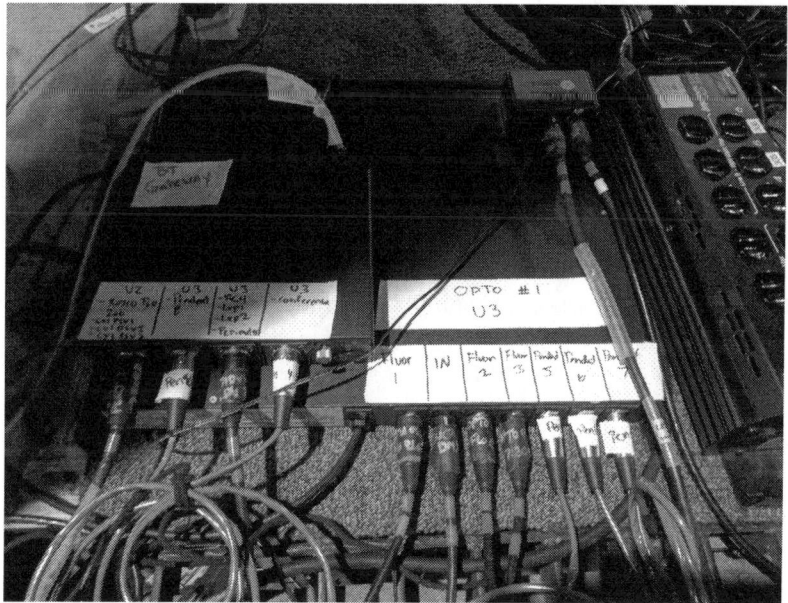

**Figure 7.21** Here a gateway is used to create four distinct universes. One universe is sent to the six-port opto-splitter so it can be duplicated across six runs.

As you plan a data path for each unit in the light plot, it is important to remember that each data run should end with DMX Termination. DMX Protocol is timed. When signal reaches the end of the run it can reflect and start to travel back the other way. This can result in signal appearing at the wrong time. Avoiding this reflection simply requires the use of DMX Termination. Many equipment manufacturers include on-board DMX termination which allows you to simply flip a switch on the last device to engage the termination. If this is not available, you can use a "DMX Terminator." A small device that looks like an XLR connector without any cable. This device contains a 120 Ohm resistor between the data + and data – pins of the connector. The resistor works to "absorb" the signal, so that it does not reflect back down the line.

## HOT POWER AND DATA PAPERWORK

Tracking all the planning around hot power and data can be difficult. As with Circuit Planning, Lightwright and Vectorworks provide some assistance with this but often it is useful to utilize other documentation too. In addition to Lightwright and Vectorworks, I use a detailed spreadsheet to give myself, the install crew, and the operator ready access to the planned system.

In Vectorworks, I set up the template with detail about the existing hot power and data infrastructures. Just like circuit infrastructure, I create a layer that includes symbols that show the locations of any dedicated or shared hot power receptacles. Similarly, you can show the termination of any permanent data runs and whether they are DMX ports or network ports. As you do the planning, draw additional symbols to denote the intended locations of temporary infrastructure equipment such as DMX gateways, opto-splitters, network switches, or circuits being used as hot power.

In some situations, it may be helpful to draw the individual cable runs in the Vectorworks file. This can help illustrate how things are connected as well as ensure the runs are of a defined length. Of course, these lines can quickly clutter your drawing, so it is important to determine if they help more than hurt. If you are interested in trying out this sort of layout, Vectorworks includes a tool for drawing cable that will also produce a list of the cable you need for the install. The Hang Plot example in the Appendix uses this tool. Knowing your cable needs is important because on big shows or in venues where no hot power or data cable is provided, you will need to determine how much cable you need to order and if that fits in your budget. Of course, there are other ways to track this data that may be more intuitive as we will see.

To set-up for data and hot power in Lightwright, go back the Dimming & Control section. In the "Universes" section of this dialog, you can set-up and label the range of universes you have in your venue. In Figure 7.22, I have them set up like I planned in Figure 7.19.

Additionally, in the "Dimmers" section you can make entries for your hot power circuits in the same way you set up your dimmers. Conveniently, you can adjust the type

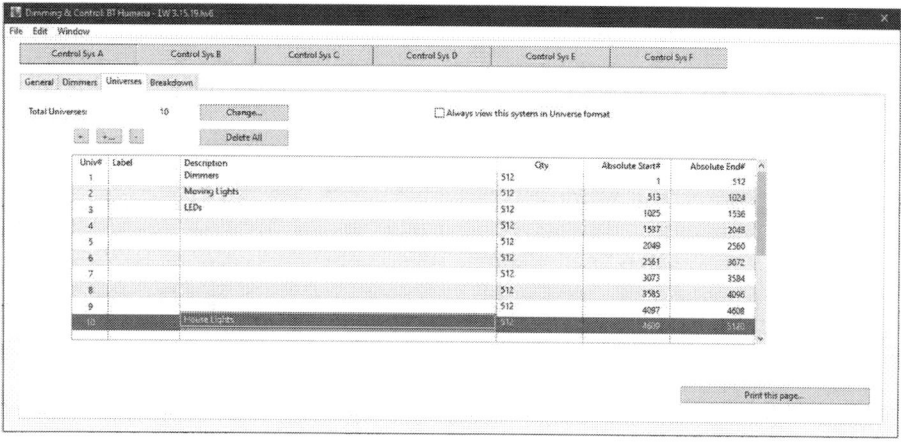

**Figure 7.22** Lightwright's universe set-up.

of these circuits to "Hot Pocket" or "ML Power" as appropriate. The other set-up area in Lightwright will be in "Instrument Type Maintenance." In this section, but sure to include a "DMX Quantity" in the "Profiles + Other" section. This DMX quantity will be the DMX Footprint for each unit. You can look up this number in the fixture's manual.

With these elements set-up, you can use Lightwright to record the hot power circuits that will be used for each fixture by simply entering that hot power circuits' name or number into the "Dimmer" column. For convenience, I alter the name of this column in the "Column Names" set-up so that it is called "Dimmer/HP." This way it is clear to the lighting technicians that the listing here may be hot power information instead of a dimmer number. Depending on the venue, I will also use a different naming convention for the hot power than used for dimmer numbers. For example, if the dimmers are numerical, the hot power might be letters. Or, perhaps, both are numerical but the hot power numbers are well out of the dimmer range. Either way will clue the installing lighting technician that something is different about this information and not go around looking for a dimmer or circuit number that does not exist.

For the data information, Lightwright has a separate column called "Address." Here you only need to enter the starting address of a unit. For example, Spiider Number 1 in our 99 Spiider example was address 1 through 123. In the address column we only need to record "1." We would set its' DMX Footprint as 123 in the Instrument Maintenance dialogue.

If the Universe section of Lightwright is properly set-up, you can also change the address format to "Universe." This will now show the address of Spiider Number 1 as 1/1—Universe 1, Address 1. This manner of displaying the universe and address is often more helpful for the installation technicians. First, this allows them to quickly see which units can be daisy-chained on the same universe if you did not otherwise specify. Second, it allows the lighting technicians to assign the fixture an address in the its' settings without doing any additional math.

For example, Spiider Number 5 in our 99 Spiider example would have to be address number 513. This address is what is called "absolute" because it does not reference a universe number. With the universe reference, the address would be written as 2/1. Since the fixture's settings do not care which universe the fixture is in, only the position of the fixture within its' universe, this representation is often more helpful. When setting up the fixture, you will assign it an address with respect to its' universe. In this case "1." It will respond to only Universe 2 commands because the cable it receives carries only Universe 2.

Finally, Lightwright can use the same error checking and report functions you saw before to check for overlapping DMX ranges, fixtures that extend outside of a universe, hot power that is overloaded, and many other potential issues.

Despite all this information there are still some other bits of information that are not readily expressed in either Lightwright or Vectorworks. For example, the elements of the hot power and data infrastructure—power supplies, gateways, and opto-splitters—are not listed as lines in Lightwright, however they often require an address and hot power. Further, there is not a good way in Lightwright to represent the order a daisy-chain should be assembled in, or which device is the end and therefore needs termination. As we have established, you are making many important decisions in your prep process, it is important to make sure those decisions are recorded and passed on clearly to your crew.

To help with this, I use a Hot Power and Data Spreadsheet like the one in Figure 7.23. Here I record all the data about my rig. Normally, when I plan hot power and data, I

**Hot Power - DMX Distribution**

Show: BT Humana 2019
Venue: Bingham Theatre  Restore Tunnel Gateway from Pipeline Location
Designer: Heather Gilbert  Ports E & F are Universe 4; Port D and B are Universe 3; Port C and A are Universe 2

| Device | Position | Unit # | Volts | Amps | Watts | HP# | HP Volt | HP Pos. | HP Pos. # | AWG/Length | DMX-512 | Uni | DMX Pos. | DMX Pos. # | Length | DMX Daisy |
|---|---|---|---|---|---|---|---|---|---|---|---|---|---|---|---|---|
| CXI PS-B1 | 1:30 BUG (Near Truss) | | 120 | 1.67 | 200 | K8 | 120 | 12:00 BUG | | Power Strip | 285 | 2 | Port C (BUG @ 4:30) | | 50 | C |
| Wybron CXI Color Fusion 7.5" | Truss 3 | 3.1 | 24 | 1.04 | 25 | | | | | | 291 | 2 | CXI PS-B1 Port 1 | | 15 + 15 | CXI-1 |
| Wybron CXI Color Fusion 7.5" | Side Pipe 2 | 1.1 | 24 | 1.04 | 25 | | | | | | 299 | 2 | CXI PS-B1 Port 2 | | 25 | CXI-2 |
| Wybron CXI Color Fusion 7.5" | 10:30 Bridge 1A | 1.1 | 24 | 1.04 | 25 | | | | | | 311 | 2 | via CXI @ SP2 | | 25 | CXI-2 |
| Wybron CXI Color Fusion 7.5" | Side Pipe 7 | 1.1 | 24 | 1.04 | 25 | | | | | | 297 | 2 | via CXI @ 10:30 1A | | 25 | CXI-2 |
| Wybron CXI Color Fusion 7.5" | 7:30 Bridge 1A | 15.1 | 24 | 1.04 | 25 | | | | | | 307 | 2 | via CXI @ SP7 | | 25 | CXI-2 |
| Wybron CXI Color Fusion 7.5" | 1:30 Bridge 1A | 15.1 | 24 | 1.04 | 25 | | | | | | 301 | 2 | CXI PS-B1 Port 3 | | 25 | CXI-3 |
| Wybron CXI Color Fusion 7.5" | Side Pipe 4 | 1.1 | 24 | 1.04 | 25 | | | | | | 293 | 2 | via CXI @ 1:30 1A | | 25 | CXI-3 |
| Wybron CXI Color Fusion 7.5" | 4:30 Bridge 1A | 1.1 | 24 | 1.04 | 25 | | | | | | 305 | 2 | via CXI @ SP4 | | 25 | CXI-3 |
| Wybron CXI Color Fusion 7.5" | Side Pipe 5 | 1.1 | 24 | 1.04 | 25 | | | | | | 295 | 2 | via CXI @ 4:30 1A | | 25 | CXI-3 |
| City Theatrical DMX Iris | Truss 4 | 4.1 | 120 | 0.05 | 6 | K8 | 120 | 12:00 BUG via CXI Power Strip | | 16/5' + 16/15' (on truss) | 453 | 2 | via CXI PS @ 1:30 BUG | | 15+15 | C |
| MB DMX II | Mirrorball Rig | 1 | 120 | 0.09 | 10.5 | K8 | 120 | 12:00 BUG via Retractor and CXI Power Strip | | 16/25' | 501 | 2 | via DMX Iris @Truss (Runs Thru Reel) | | 15+25 +3p Adapter | C |

Figure 7.23 A typical hot power and data sheet.

do it using this sheet first and then copy the data from the sheet to Lightwright and Vectorworks for installation use and error checking. The Hot Power and Data Sheet includes a line for each item in the plot—both designed fixtures and infrastructure elements. Items are listed in the order that they connect to each other so that daisy-chains are easily seen.

For example, color scrollers are normally listed in the Lightwright, but their power supplies are not. Their power supplies will still have a significant power draw to consider as well as an address. On the Hot Power and Data Sheet shown here, I have made a line for the power supply and then underneath are lines for each scroller that connect to it. This clearly shows how those bits of equipment relate to each other and provides a quick reference for the important data about each piece.

# REFERENCE

Bennette, Adam. 2008. Recommended Practice for DMX512: A Guide for Users and Installers. PLASA/USITT.

# CHAPTER 8

# Documentation and Shop Prep

## LOAD-IN DOCUMENTATION

Once you have thought through your electrical and control plans, you complete the Paperwork Prep process by preparing the actual documentation that your installation team will use to do their work. If you have been diligent in your plot clean-up and electrical planning to this point, a good bit of this work will be already done. I find it helpful to think of the planning process in general as a long road toward this final paperwork. The initial Lighting Design submission details the "what" of the Lighting Plot, but you are creating documentation that describes the "how." For example, when I am entering circuit information into Lightwright or when I am entering daisy-chain information into the Hot Power and Data Sheet, I am not doing it just for me or because I am simply filling out a form. I am making a road map for the installation team. I am thinking through all the information that they will need to complete their work. Where does this light plug in? What kind of cable will I need? How does that moving light receive its' data? Making decisions in advance is incredibly important, but it will not matter if you cannot easily convey those decisions to your team.

The complete paperwork package for the installation—also called "Load-in" or sometimes "Changeover" when the previous show immediately transitions to the next—includes the following:

- Master hang plot.
- Hang cards or hang tapes.
- Load-in Instrument Schedule.

- Mult Sheet
- Hot Power and Data Sheet.
- Color and template count sheets.
- Color and template prep sheets.
- Rigging installation documentation.

I strongly recommend that all load-in documentation be printed in hard copy. You should keep a full set of current paperwork in a dedicated three-ring binder. This copy of the paperwork should be considered the master copy. Never hand out your master paperwork to the installation technicians. Print additional copies of that paperwork for them so that they can have their own copy to use with impunity.

Today there is a push to "go paperless." In general, I am in favor of this. However, I never use a computer or tablet as a replacement for my master paperwork in the field. I do not want to risk wasting time if the computer hangs or crashes when I need to look up an answer for a technician. Additionally, I would not want to subject a computer or tablet to the dangers of being dropped from height. I have seen countless Instrument Schedules fall from catwalks. If I had to choose, I would much rather see someone drop a packet of paper or a three-ring binder than a tablet.

Further, regardless of how well you plan there will likely need to be some changes in the field. I find it helpful to have a clear indication of what those changes were when I review the installation's progress at the end of each workday. To track this, I have technicians record changes directly into my master paperwork. Then, I have them fold over the corner of each page that contains a change. At the end of each day, I reconcile my master paperwork with my digital copy. After reconciliation, I highlight each change in the hard copy and unfold the corner of the page to notate that the change was recorded in the digital paperwork. Manually reconciling changes like this at the end of the day gives you an opportunity to consider their impact in greater detail. While most computer systems give you an ability to track changes, I find that it is much easier to see and evaluate changes when going through this manual reconciliation process.

The first piece of paperwork you will generate is the Master Hang Plot. This is the plot that you have been working on since the initial design submission. From the point where you started the clean-up process through all the infrastructure notes, you have been manipulating the initial design submission to transform it into a master technical schematic of the installation. This is the plot that you will carry into the venue. You want to make sure it has all the information you might need should any questions arise.

In addition to all of the circuiting, hot power, and data infrastructure information, the Master Hang Plot should include for each fixture a graphical symbol easily recognized by your crew, the fixture's channel, its' dimmer or hot power name or number, its' circuit number, its' starting address, its' color and template information, and any notes about how

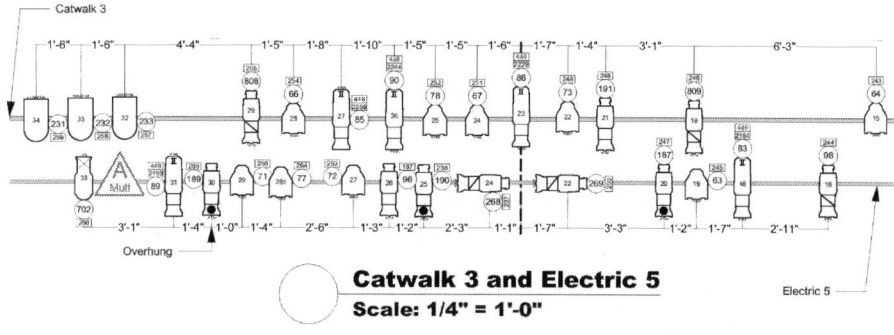

**Figure 8.1** A Hang Card made with Vectorworks showing many of the details also found on a Hang Plot.

to hang it. The hang plot should be in scale, but since having a scale ruler in the field may be undesirable, all distances should be dimensioned unless they conform to a standard spacing. For example, most light plots feature units with 18 inches between them. If that is the case for your plot, you can use a note that says, "all units with 18 inch spacing unless noted." Then you only need to dimension the units that do not fit that spacing.

Additionally, it is best to always dimension your plot with reference to the center of each position rather than the ends. While many venue drawings are reasonably correct, the exact length of the pipes used is often not accurately represented in the drafting. To ensure that lights are placed in the correct place, always measure from the center of a pipe.

As a full light plot is large and often difficult to hold in the field, only one copy of the Master Hang Plot is normally is required. The Lighting Supervisor should keep it in a safe place for reference. In lieu of individual copies of the plot for each technician, the Lighting Supervisor can produce "Hang Cards." Hang Cards are smaller subsets of the full hang plot that show only one position or small group of positions. These are designed to be easily held by the installation team as they work. Originally, hang cards were made by taking a secondary copy of the plot, cutting in with scissors, and adhering the smaller pieces to cardboard. This created a sturdy reference drawing that could withstand the rigors of the install.

As the CAD skills of Lighting Supervisors have dramatically improved over the last decade, many of these handmade hang cards have been replaced by dedicated hang drawings made in the Vectorworks or other CAD software. In Vectorworks, you can use the "viewport" feature to "cut out" the appropriate section of the plot and then "paste" it onto a sheet layer. This digital process is a perfect analog to the traditional method process of preparing hang cards, but it provides much greater flexibility and efficiency. For example, you can easily print out any number of hang "cards", so it no longer matters if the "card" is sturdy enough to survive the field. If it is ruined, you can have another copy floating around.

**Figure 8.2** A Hang Tape printed with the "Lightning Tapes" Plug-in.
Courtesy of Christopher Hetherington and Lightning Tapes.

In some venues, particularly when working with positions that can be flown in, the Lighting Supervisor can make "Hang Tapes" or "Truss Tapes" instead of hang cards. A hang tape is a long strip of paper—usually cash register paper—that includes the details of each light spaced in life-sized scale. This allows the installing technicians to stretch the tape across a position and see exactly where each light should go as well as have ready access to all its' details—circuit number, dimmer number, color, and so on. Like hang cards, hang tapes were originally made by hand. The Lighting Supervisor or their team would copy the plot data from the plot to the cash register tape with a marker.

Technology has brought greater efficiency to this process as well. In Lightwright, for example, you can print your fixture information on standard shipping labels and stick them to the hang tape as you measure. This allows the information to be very detailed without you needing to spend time writing it out. A couple of enterprising technicians even went so far as to develop a software plug-in for Vectorworks called "Lightning Tapes." This plug-in uses Vectorworks' data and a cash register tape printer to print the information directly on the tape in full scale.

In some special cases other methods of plot documentation can speed up the process of translating information to the installation technicians. When lighting positions are easily accessed during the run of the previous show in a venue—as is the case in many catwalk or tension grid venues—the Lighting Supervisor can run a "Dry Hang." In a Dry Hang, the Lighting Supervisor will read off the position measurements to a technician in the grid and that technician will adhere a label—like the ones used in making Hang Tapes—directly on the pipe. With these labels, all the information for each light is attached to their position and it is not necessary for the installing technicians to carry any additional paperwork. This is particularly helpful in venues where access to the positions is challenging. It allows the technicians in these situations to focus on the equipment and themselves over paperwork.

In other venues, a Repertory Numbering System may be used. Here the positions will be marked with regular tick marks—usually at 18-inch intervals—and numbered. These tick marks are also drawn on the venue's template and all hang paperwork. When a light is hung at one of these positions, the positions repertory number is used as the light's unit number. If a light needs to be hung between tick marks, the unit number can have a letter or decimal number added to it to designate a "half position." For example, a light that is drawn at tick mark number 10, would be unit number 10 even if it were the only light on

**Figure 8.3** Repertory Position Numbers drawn on a pipe.

| Unit# | B/B | Inst Type & Watt | Purpose | Color | Gobo | Acc | Mu | Le | Cbl | Dm/H | Addr | Chan |
|---|---|---|---|---|---|---|---|---|---|---|---|---|
| **8 Electric** | | | | | | | | | | | | |
| 9z | | ETC S4-PAR MFL 750w | OMSOL DOWN | R163 | | 750w Lamp | | | 2Fer w/ 8E #10z | 297 | 1/297 | (70) |
| 10z | | ETC S4-PAR MFL 750w | OMSOL DOWN | R163 | | 750w Lamp | | | 2Fer w/ 8E #9z | 297 | 1/297 | (70) |
| 11z | SH | ETC Lustr2-36 171w | OMSL BAK -> | R163 | | | | | | 453 | 5/316 | (88) |
| 12 | | ETC S4-PAR MFL 750w | MSOL DOWN | R163 | | 750w Lamp | | | | 301 | 1/301 | (69) |
| 13 | SH | ETC Lustr2-36 171w | OUSL BAK -> | R163 | | | | | | 453 | 5/331 | (92) |
| 14 | | ETC S4-PAR MFL 750w | OUUL DOWN | R163 | | 750w Lamp | | | 2Fer w/ B5 #9 | 300 | 1/300 | (80) |
| 26 | | ETC S4-26 575w | SPARE | R119 | | | Z | 5 | | 335 | 1/335 | (812) |
| 27 | | Altman 6x16 750w | PICTURE FRAME | N/C | | | Z | 6 | | 336 | 1/336 | (713) |
| 29 | | Altman 6x16 750w | PICTURE FRAME | N/C | | | Z | 7 | | 337 | 1/337 | (712) |
| 29z | | Altman 6x9 750w | | R119 | | | Z | 8 | | 338 | 1/338 | (106) |
| 31z | | Altman 6x16 750w | PICTURE FRAME | N/C | | | Z | 9 | | 339 | 1/339 | (711) |

**Figure 8.4** A typical instrument schedule.

that position. If the light were to be hung between tick mark 10 and tick mark 11, it could have the unit number of 10a, 10.5, or any other designation that is easily understood by the installation crew.

One of the big advantages of the Repertory Numbering System is that the Instrument Schedule can be used in place of a hang card or other documentation. The Instrument Schedule—normally generated by Lightwright—is an important piece of paperwork because it details all the information about each light on a position in the order it appears on the position. Because the Instrument Schedule is a spreadsheet instead of a drawing, it can display much more detailed information. Consequently, even if the installation technicians have hang cards or hang tapes, it is still good practice to provide them with copies of the Instrument Schedule. This additional paperwork gives the technicians a cross-reference if additional clarity is needed. For example, I always include a special column in my Instrument Schedules with notes about how to run power for potentially confusing lights. This way, if I need to use a circuit far from a light, I do not have to worry about the technicians spending valuable time looking for that circuit. They simply consult their Instrument Schedule and know where to go.

## COLOR AND TEMPLATE PREP

In a typical schedule, Color and Template Prep occurs prior to the rest of Shop Prep. Despite being a shop activity, this process is essential to the Price Out and therefore is best completed during that time. Some commercial producers discard all color and templates at the close of each production. Color and Template Prep for these productions is easy because all needed expendables are purchases. No stock checking is required. However, most producers keep a color and template stock from production to production until it is no longer reusable. With these producers, the Lighting Supervisor will have to evaluate how much of the color and template stock can be used in a particular show before purchasing more.

The Color and Template Prep Process that I use has two phases—counting and framing. The first phase—counting—can happen almost as soon as the plot is submitted. I usually start the counting phase right after completing my first plot clean-up pass so that I can have the results of the counting available right away.

Lightwright is especially useful in the counting phase as it can give you or the technicians who will count your inventory a clear total of what to look for. In the software's "Instrument Type Maintenance" section, each instrument can be assigned a "Color Frame Label" and a "Gobo Size." As part of your template file, ensure that each instrument in your inventory is listed in this maintenance and that each one has data in these columns as appropriate. When you are doing your plot clean-up, be sure to check that this data is assigned appropriately. If your plot is using fixtures beyond your inventory, you will need to add this data.

For "Color Frame Label," record the size needed for the instrument. For example, an ETC Source Four with a 36-degree Lens Tube takes a 6.25-inch cut of color—or "gel." The label for that fixture would read "6.25 x 6.25" to show that that fixture needs a square 6.25-inch cut. Assigning labels like this makes your Color Count Sheet much easier to read and understand.

It also helps take much of the guessing out of process for any technician you assign to count color. If you did not assign labels, for example, Lightwright would generate a count sheet that says that you need *n* number of cuts for a 36-degree Source Four. Since no size is listed, the technician would need to know what size that fixture takes.

Additionally, many lights take the same size cut of gel. If you assign Color Frame Labels to all your instruments, Lightwright will combine those entries. Instead of having a long list that asks for 3 cuts for a 36-degree Source Four, 10 cuts for a 26-degree Source Four, and 17 cuts for a 19-degree Source Four, you can have a simple print out that says that you need 30 cuts in a 6.25 x 6.25 size. You or the technician can simply count out those 30 and move on without caring what they go to.

For templates, a little more leg work is required. In the "Gobo Size" column for each fixture, I will enter the largest size template that the fixture can use. Many lights can only use one size template, but some—like the ETC Source Four just mentioned—can take more than one size. As with color, I set up the Gobo Size as part of my template file and then check it during the plot clean-up.

Unlike color, I do one more step with templates during the plot clean-up as an added check. For each fixture entry in the main Lightwright worksheet, I will check to see if the size of the template is specified by the designer in the fixture's gobo column. Often designers will only specify the pattern, but not the size. Most of the time designers will prefer the largest size they can use, but if there is any uncertainty, I will red flag it early on and ask them about it.

Once I confirm the designer's desired template sizes, I will change each template entry in the Lightwright worksheet to reflect both the size and the pattern. For example, a Rosco pattern 77777 is normally written "R77777." If I determine that the designer wants a unit to use the "Size A" version of this pattern, I will change the entry to "R77777-A."

However, If the fixture in question were an Altman 360Q, I would need away to know that "Size A" is too big for that unit type. Since I would have set the fixtures' default Gobo Size to "B" in the Instrument Schedule Maintenance, when I print out my Template Count Sheet, it will say that I need $n$ number of B Sized R77777-As. As that does not make much sense, I will immediately catch that there is some sort of error. In this case it is that the designer is specifying a gobo size that they cannot use.

Once all the color and template information is set-up in Lightwright, I print the Color and Template Count Sheets to inventory the stock. Once printed, I find it helpful to make each printout into a form like the one shown in Figure 8.5. For Color Count Sheets like this one I make four columns—Need, Pulled, Sheets, and Order.

The "Need" column is the typed information from Lightwright. This will be total of a particular color and size used in the production.

| COLOR COUNT | NEED | PULLED | SHEETS | ORDER |
|---|---|---|---|---|
| **Color Cuts** | | | | |
| L161 | | | | |
| 7.5" size, 7.5" X 7.5" | 1 cut | 2 | 0 | — |
| Total: | 1 cut | | | |
| L200 | | | | |
| 7.5" size, 7.5" X 7.5" | 50 cuts | 75 | | |
| S4 size, 6.25" X 6.25" | 3 cuts | 5 | 2 | — |
| Beam Projector size, 14" diameter | 1 cut | 2 | | |
| Total: | 54 cuts | | | |
| L201 | | | | |
| 7.5" size, 7.5" X 7.5" | 34 cuts | 20 | | |
| S4 size, 6.25" X 6.25" | 5 cuts | 8 | 0 | 5 SHEETS |
| 5K Large size, 16" diameter | 1 cut | 2 | | |
| Total: | 40 cuts | | | |
| L202 | | | | |
| 7.5" size, 7.5" X 7.5" | 42 cuts | 63 | 1 | 1 SHEET |
| S4 size, 6.25" X 6.25" | 6 cuts | 0 | | |
| Total: | 48 cuts | | | |

Figure 8.5 A Color Count Sheet made into a hand-written form for stock checking.

The "Pulled" column is the number of cuts that are able to be pulled from our stock. I always attempt to pull 150% of the need count so that I know how much spare color is available for the run. This means that the "Pulled" number will be higher than the "Need" number unless the amount available is not enough to provide ample spare.

In the "Sheets" column I record the stock number of uncut sheets of that color. Gel is normally sold in sheets and rolls. It is best to only cut it when it is going into a show so that you have flexibility on the sizes of your spare color. The "Sheets" number is important if any of the "Pulled" numbers for a color are lower than 150% of the need numbers. As I consider how many sheets I need to get to my ideal pull number, I also should consider how many sheets I have.

Finally, the "Order" column is where I will write my needs after evaluating the numbers from the count. One I have completed my count and ordered the color, I always save this sheet as a record of what I needed. From time to time, I have accidentally ordered the wrong amount of color. Keeping this sheet lets me check the invoice or packing slip against it when the order arrives. This way, I can recognize any mistake early in the process and immediately place a second order.

For the Template Count Sheet, I use a similar method as seen in Figure 8.6. However, in this case my columns can vary depending on the venue. Template costs can add up quickly, so there is great advantage to using stock templates if they are in good condition. Like all expendables, templates eventually wear out, but many patterns can last for years if

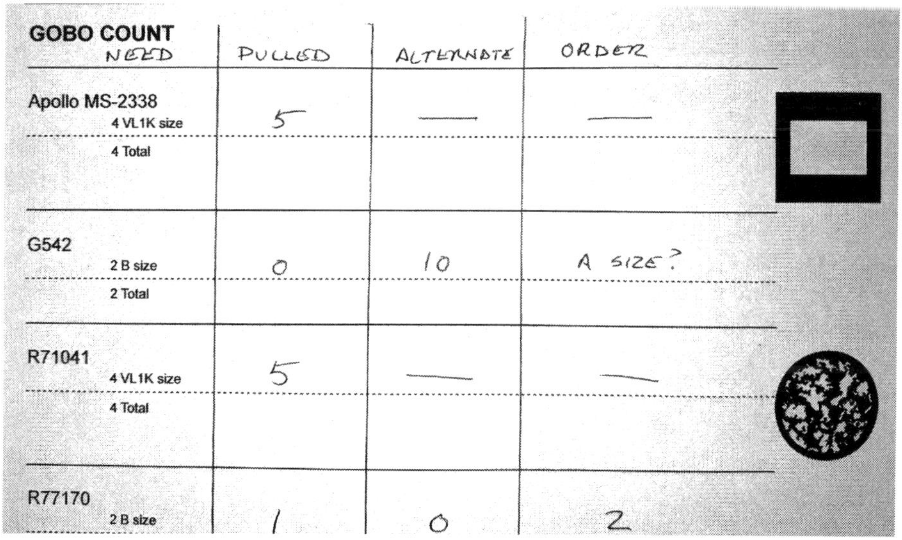

| GOBO COUNT | NEED | PULLED | ALTERNATE | ORDER | |
|---|---|---|---|---|---|
| Apollo MS-2338 4 VL1K size | 4 Total | 5 | — | — | |
| G542 2 B size | 2 Total | 0 | 10 | A SIZE? | |
| R71041 4 VL1K size | 4 Total | 5 | — | — | |
| R77170 2 B size | | 1 | 0 | 2 | |

**Figure 8.6** A Template Count Sheet made into a hand-written form for stock checking.

not heavily used. To help steer the use of stock templates, I always count both the designed template size and any viable alternates. If I can fill the need with an alternate size and I need to save some money in the budget, I will ask the Lighting Designer if an alternate size will be a problem. Most of the time, designers have some flexibility with gobo size, but there will be times when the design absolutely requires a specific size. Like any other element of the design, you will have to prioritize that need with the Lighting Designer and determine how it fits in the budget.

To understand where the stock template inventory stands, I use four different columns on my hand made form—Need, Pulled, Alternate, and Order.

Like the Color Count Sheet, the "Need" column is the printed numbers from Lightwright and "Pulled" is the number pulled for the production. For templates, I normally pull 120% or one spare gobo for every five active gobos.

Instead of "Sheets," I use a column labeled "Alternate." If the inventory has the same pattern in a different size, I can record information about those sizes here. For example, if the "Need" column required $n$ number of A Size templates, I would record the number of those templates I was able to pull in the "Pulled" column and then record any B Size templates still in stock in the "Alternate" column. This way, if the "Pulled" column number is less than the "Need" column number, but the "Alternate" column number exceeds it, I can save money by recommending that the size be changed if possible.

Finally, the "Order" column is used in the same way as in the Color Count Sheet.

With the counting phase complete, you can move to the framing phase. For color, first go back to the Instrument Type Maintenance screen in Lightwright to make any adjustments for fixtures that have special frame requirements. For example, the Altman 360Q and the ETC Source Four PAR both use 7.5-inch square gel cuts. At the beginning of the process, you would have set their "Color Frame Label" to "7.5 x 7.5" to allow for their counts to be combined. However, as you look to frame those cuts, you know that ETC Source Four PAR color frames feature a larger aperture than Altman 360Q frames, so you need to change the Color Frame Label to reflect that. This now allows the different frames to be counted separately.

Next, print out the Color Count Sheet again, but this time instead of using the default "Count as Color Cuts," select "Count as Framed Color." This new sheet—like the one in Figure 8.7—will provide a full count of which color or colors go in which frame size. For example, the Color Count Sheet used before listed $n$ cuts of Color $y$. In this new sheet, it will tell you that you need to prepare $n$ frames of Color $y$ and $x$. This allows you to prepare stacks of framed color for the show.

You can do the same for templates. For this printout, change the "Gobo Size" field in the Instrument Type Maintenance to reflect the type of template holder you will need. For example, ETC Source Fours will need ETC Source Four template holders. For those units, you can list "S4" in the Gobo Size field. Since you will have recorded all of the

### COLOR COUNT
#### Framed Color

● L161+R119

| | |
|---|---|
| Altman size, 7.5" X 7.5" | 1 frame |
| Total: | 1 frame |

● L200

| | |
|---|---|
| S4 size, 6.25" X 6.25" | 3 frames |
| Beam Projector size, 14" diameter | 1 frame |
| S4 PAR size, 7.5" X 7.5" | 15 frames |
| Altman size, 7.5" X 7.5" | 4 frames |
| Total: | 23 frames |

● L200+R114

| | |
|---|---|
| Altman size, 7.5" X 7.5" | 1 frame |
| Total: | 1 frame |

● L200+R119

| | |
|---|---|
| Altman size, 7.5" X 7.5" | 30 frames |
| Total: | 30 frames |

**Figure 8.7** A typical Color Framing Sheet.

actual gobos sizes as part of the individual worksheet entries in the previous step, the resulting print out will tell you that you need *n* number of a certain gobo size and pattern in a "S4" frame. When printing out this sheet, use the same settings you used for the first template count sheet.

With both count sheets ready, you can start the framing process. Going line by line, you or any technicians can put the color and templates in frames as required for the show. As this process has no impact on any work being done in the venue, it is ideal to do this work before the installation window, usually while any previous show in the venue is still in performance. As this is the case, it is usually recommended to have more color frames and template holders than needed for your full inventory thus allowing you to have an "active" frame in the light and a "prepped" frame in the shop for the next show. If that is not possible and you run out of frames, I recommend using pieces of scrap paper or paper clips so that the sorting work can still occur and transfer to proper frames after the previous show strikes. Even if the color and templates need to be transferred to frames later, time will be saved by having everything organized.

Depending on your venue, you may wish to end your color and template prep process there. However, if you are working in a nongrid and/or nonladder venue, it is helpful to organize your frames by position. If you are not sure whether this would be helpful, I would recommend trying it. This process is helpful most of the time. When you encounter a situation that it is not, it will become clear what situations make it difficult to do this in.

To organize your frames by position, start by printing a copy of the Instrument Schedule. Going down the list, pull frames from your preframed stacks and set them in the order they appear in the schedule. If you encounter a fixture that has no color, insert some sort of placeholder for it. Normally, I use an empty frame because it allows me to give that unit a frame anyway. Now if that unit needs color added during the tech process it already has a frame. If you have a fixture that requires a template, put the template holder, with the template in it, inside the color frame with the color. This designates that that template and that color belong to the same fixture.

When the color and template stack for a position is finished, bind it together with rubber bands or tie line and slip that position's Instrument Schedule in front of the first frame to denote which side of the stack is the lowest unit number. Set aside these Color and Template stacks until the plot is hung. When the install is complete, the installation technicians can quickly run through and insert the appropriate frames from the stack.

## RIGGING PLANNING AND PAPERWORK

In addition to your work in electrical planning, the Lighting Supervisor is also responsible for the structural soundness of the lighting plot. In most purpose-built venues, fixtures will be installed on dedicated rigging infrastructure. However, even if all positions used in the light plot were permanently installed by outside contractors, you still must understand their structural limitations. No lighting position will be able to hold an infinite amount of lighting equipment.

Much like evaluating your electrical infrastructure, you need to evaluate the rigging infrastructure of your venue. You will want to evaluate each position to understand what factors will influence its' maximum point loads—which can vary over the length of the position—and the factors that will influence the overall maximum load when that load it evenly distributed. If working in a venue with a counterweight system, you will need to determine how much counterweight each position needs and how to safely install that weight during the load-in. Finally, if you are working in a venue without a permanent infrastructure or if you need to add additional positions to achieve angles not possible in your venue's permanent infrastructure, you will need to evaluate the building itself and the impact adding new positions will have.

Planning a rigging infrastructure or evaluating an existing one is a complicated process. Fundamentally, you are doing the same thing you did when you evaluated for electrical compliance. You will need to determine the capacity of the system components to find your weak spots and then determine if the actual load of the light plot exceeds that capacity. The reality, of course, is that the process of evaluating structure can be quite complex. Indeed, many books have been written solely on this subject. For those beginning their careers as Lighting Supervisors or Lighting Technicians, I would strongly encourage additional exploration of some of these texts. I have listed several of these resources in Chapter 14.

Additionally, hands on training and consultation with your venue's Technical Director will certainly assist you as well. Almost every venue has certain quirks to it, consulting with colleagues that have experience in the venue can help illuminate those quirks.

With time and effort, you should be able to develop the rigging expertise needed to evaluate these systems on your own, but until then remember that it is not essential that you be the evaluator, but that the systems are evaluated. Use the expertise of colleagues and professionals in this area to ensure that the work being done is safe and structurally sound. Remember, the Lighting Designer's creativity will push your venue to its' limits. It is up to you to know what those limits are.

For example, standard hanging positions are made with 1-1/2 Inch Schedule 40 Black Steel Pipe. This pipe weighs 2.72 pounds per foot, ETC Source Four 36-degree Fixtures weigh 14 pounds, and 12 A.W.G. SOOW Cable weighs roughly 0.25 pounds per foot. If you had a 20-foot span with 10 of these fixtures each with 20 feet of cable, your pipe would weigh about 200 pounds. In order to ensure that the pipe can withstand this weight, you will need to evaluate the capacity for the aircraft cable holding it up, any inline hardware, the ceiling superstructure, and, if it is a counterweight position, the capacity of the arbor.

Recognizing that the installation of your light plot will require this structural evaluation, it is important to document the manner in which the plot is to be installed as clearly as possible in order to ensure that the installation lighting technicians have all the information they need to be able to do so safely. For example, if you assign a team of lighting technicians to install a boom, but do not give them pipe length, ballast, or guy wire instructions, the technicians may not install the boom in the safe manner that you have planned. If the boom collapses, you are responsible for that failure. In this, as in all things, planning is only as valuable as your ability to communicate it.

For rigging projects that require the addition of positions, it is best to produce separate paperwork. Each project will benefit from a clear drawing and a supply list. The supply list is important because it allows you to track the hardware and tools you will need for the installation to ensure that they are included in your price out. Additionally, it allows the installing lighting technicians to gather all that they need before beginning a project. This will allow them to be more efficient as well as give them a check on their work. If a

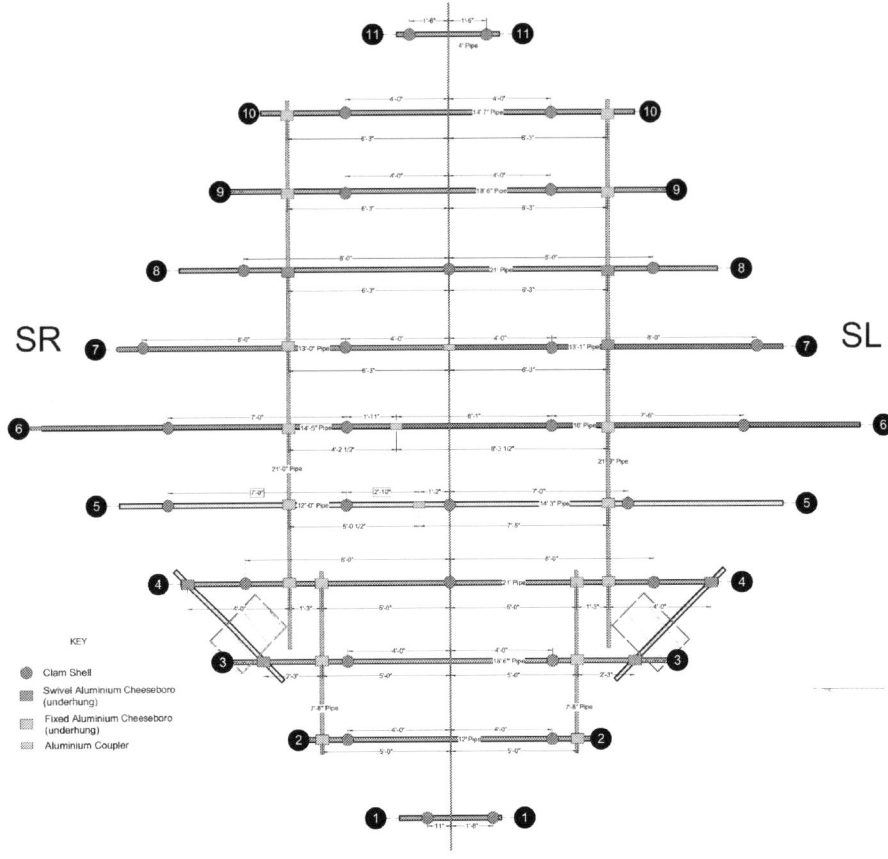

**Figure 8.8** An example rigging drawing for a temporary pipe grid installed below the permanent pipe grid.

Courtesy of Dani Clifford.

project requires ten things, for example, and at the completion of it, there was one remaining piece of hardware, it could be assumed that there was a problem with the installation worth reviewing.

Any rigging drawing should be shown both with and without lighting equipment and be fully dimensioned. The lighting equipment can clutter the drawing and make it hard to see certain elements. Ensure that all attachment points are drawn as accurately as possible. When creating the drawing, you are virtually completing the installation. The more detail you put into the process of creating the drawing the more likely you are to catch any issues that the team may run into. Think through each step and provide the notes that you would need to make sure it happens as if you were installing it.

## SHOP PREP

Once all the paperwork is completed, you will move on to Shop Prep or Physical Prep. Shop Prep normally consists of the following activities:

- Color and Template Prep
- Equipment Pulling, Labeling, and Addressing
- Pre-rigging
- Preparing Load-in Kits
- Set Electric and Practical Construction

For the Lighting Department, most of the work occurs in the brief installation period. However, it is a common mistake to assume that all the work should be done in that period. In reality, there are many things that can normally be done outside the installation window. I always evaluate each step of the load-in process and consider what, if anything, can be done in advance.

For example, consider how much easier it is to set addresses or other settings on moving lights one afternoon in a well-lit shop or rental house versus trying to do so on an overnight call with hammers banging in the background. Every production is different, and while I am making some suggestions here, it is important to always ask yourself for each production that you do, "is there anything else I can do in advance?" My motto is, "If you can prep it, you should prep it."

Since we have already discussed Color and Template Prep, I will move on to Equipment Pulling. The amount of this you do will be dependent on your situation. If you have a fully conventional rig and all your lights are already hanging from the previous show, there is not much you can do. However, if you are doing a big rental or implementing equipment that is otherwise sitting on the shelf, get it ready.

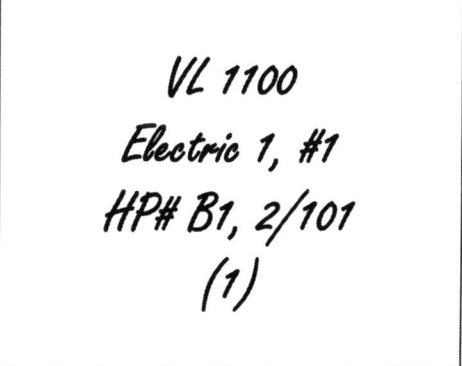

**Figure 8.9** An example of the content of a fixture label.

For in-house equipment, this process is straight forward. First, pull the gear out of storage and ensure that it works like it is supposed to. It is easier to fix something in prep than in the middle of load-in. If it needs attachments or other items, take care of that at this point. For example, if you are using a color scroller and it needs a certain string, this is the time to install it.

Second, label the equipment. The goal here is to make sure that when someone picks it up later, they know exactly where it goes and what it is for. A good label should include:

- Any equipment reference name or number—such as "Power Supply A" or "Opto-Splitter 1"
- Position Name.
- Unit Number.
- Hot Power and DMX Address, if appropriate.
- Channel Number, if applicable—for troubleshooting post-installation.

Finally, if the unit has any onboard settings—including DMX address settings—these should all be set during this time. Remember to read the manual for all equipment that will be in your light plot. You will frequently encounter equipment that is new to you because new equipment is always being made. Never assume that you know how new gear works out of the box. For example, you may have an LED fixture that you have never used before and not know that it has dimmer curve settings. Perhaps the default is "fast," but the designer is used to it working in "incandescent emulation mode." You might get to tech and discover that you must change the settings for all your fixtures before rehearsal can start. Now everyone is waiting as you move your ladder from fixture to fixture changing the settings. A simple problem that could be avoided with an adequate prep process.

If you have a large amount of in-house equipment, you can explore using default settings to save time. For example, you can determine that all your moving lights will always be in the second universe and give them default addresses that do not overlap with each other. Now, when you pull out those lights for prep, they already have their addresses set.

If the equipment you are using will be rented, most rental houses will allow you to do this work in their shop prior to the rental delivery. This is especially important when you have a large rental package and not a lot of space at your venue.

Next, we move on to Pre-Rigging. As discussed in the Rigging Paperwork section, you will prepare a supply list for each rigging project. Many of those supplies can be pre-assembled. For example, if a new position will be hung from five custom length aircraft cable points, you can cut and swage each of those points in advance of load-in. Then at load-in you are simply connecting pieces together.

For both rigging projects and general load-in projects, you can prepare Load-in Kits. Using plastic bins, hampers, road cases, or other storage containers, combine equipment and tools to make kits for various projects. For example, if you are installing a boom, you might want to have a hamper with the boom base, additional weights, side arms, and tees ready to go. Now when you want to install that boom, you can hand off the appropriate paperwork and all the supplies allowing the technician to get started right away. They do not need to waste installation time looking for supplies.

The final element of Shop Prep is Set Electric and Practical Construction. This will vary depending on each show you do, but invariably will occur with some frequency. Set Electrics or Practicals are any elements of another design area that requires lighting involvement. This could be an LED tape halo under the edge of the stage, a streetlamp, or a light-up hat. Regardless, try to coordinate time with other departments to work on these projects outside of the installation. Often these projects take time and concentration that is difficult to find during the intense installation window. Even if the entire project cannot be completed prior to installation, it is good to find some elements that can.

In addition to setting aside time to work on these projects, be sure to generate detailed documentation about how they are to be constructed. In many cases you are building custom designed fixtures from component materials. Be sure to start planning these projects as soon as they come up. Often these ideas are discussed in design meetings prior to the submission of the light plot. There is no reason to wait until the light plot is submitted to develop a schematic plan for Set Electrics or Practicals. This plan will help you when you are developing your Price Out or working with other departments in the early budgeting stages. When you reach the time to execute the plan, the schematics will be helpful conveying your plan to your technicians. More information on Set Electrics and Practicals is in Chapter 12.

# CHAPTER 9

# The Load-In

## DANCING NOT FIGHTING

When the time comes to move into the venue, the planning and scheduling you have done is put into action. Load-in is a very tense and fast-paced process. Previously, you were thinking was in days, but now you need to think in hours. All the technicians in your organization are fighting over the same 50 cubic feet. If you have planned and collaborated well, you will be in good shape, but you must remain adaptable as even the best plans will need to change as unforeseen issues arise.

The first thing to remember is that your schedule does not exist in a vacuum. Every other department has their own schedule. Before load-in begins you will need to compare your schedule with those of others and negotiate conflicts. For example, you cannot set up booms if the scenery department has removed the floor. In an ideal world, these schedules intertwine like a carefully choreographed dance. In a worst-case scenario, you end up with a fight for space and time. When planning, you want to be flexible, but also advocate for your needs. Be sure the things you advocate for are essential things. If you can find a way to shift, you should be willing to change. In many cases, Lighting Supervisors are resident staff members at an organization. The Technical Director you argue with on show number one, will still need to be your ally on show number five. Play the long game.

When discussing the schedule with other departments I consider each project—theirs and mine—like a Tetris piece. Assigning a schedule slot to a piece generates an impact on the other pieces. All department managers need to work together to fit the pieces in the master load-in plan. Of course, the real challenge comes when everyone is in the venue. Some projects take longer than expected and the shape of each Tetris piece morphs. In these moments, the detail of the schedule is what can save it. As one piece morphs, the way in which the other pieces can respond is evident only if those pieces are appropriately detailed.

As an example, consider a schedule where "Lighting Load-in" is one large piece and "Scenic Load-in" is another large piece. If either piece unexpectedly grew, the other would not fit in the puzzle anymore. Whereas, if the Scenic piece is divided into "Deck Install," "Wall Install," and "Paint" and the Lighting piece is divided into "Hang," "Cable," and "Set Electrics," it is easier to slide them around each other. The more detail each piece has the easier it is to evaluate the impact of adjustments to other pieces.

If you look back at Figure 6.5 from Chapter 6, you can see a great example of these puzzle pieces. In that schedule, Tuesday the 11th has some members of the Lighting Team installing LEDs into lightboxes. This was planned because the Lighting Department was scheduled to receive lightboxes from the Scenery Department the day before. If the Scenery Department is a day late, what is the impact of that on the Lighting Department's schedule? Should they take the day off and push everything down a day? Should they move something from later in the week to Tuesday so that they can work on it while they wait for the lightboxes? The detail of your schedule and the schedule of the Scenic Department will help you decide what to choose.

To help shifting these puzzle pieces, I recommend including some "catch-all" or "contingency" time. The schedule from Figure 6.5 shows multiple days of "catch-all" time leading up to focus. Given the amount of contingency time available, it might make sense to simply push back some of Tuesday's work to Wednesday and readjust. This would allow you to give some of the technicians time off on Tuesday which could greatly help morale. I love contingency days. More often than not they become a day to rest and take a breath before Focus, but even if you have to use them, they help keep your stress level down.

Remember that if you use your contingency days or change your schedule in some other way it is important to discuss those changes with the other department managers. Even a quick e-mail or phone call goes a long way. It is easy to get ahead and plow on through, but this might hurt another department if they are not given the heads up. For example, if you needed the full stage for ladder work on day three, but were ready for it on day two, taking the stage on day two could hinder the on-stage paint work that the Scenic Department planned to do that day and put them a day behind.

## INSTALLATION BEST PRACTICE

When approaching the load-in work, the Lighting Supervisor must remember that it is their responsibility to define and enforce quality and standards among the lighting staff. The Lighting Supervisor sets the tone for the installation. As any good technician knows, unless it is unsafe, the house way is the best way. It is, therefore, up to the Lighting Supervisor to set clearly defined standards and practices for the technicians working in their house.

When the technicians are on full-time staff with the Lighting Supervisor, this process is more straightforward. I recommend recording the standards and practices you wish to use in a department in a handbook to share with your staff. However, if your staff is largely temporary, this level of detail is often difficult to manage. For these situations, I recommend preparing a simple one sheet that you can read to technicians at the top of the workday or send to them in advance. In either situation, you should be mindful of the work being done and try to catch standards or quality deviations early so that a larger percentage of the work is done correctly.

To help with quality and efficiency, I recommend that most work be done in waves. This is not always ideal, as I will explain shortly, but in general, working in waves allows for different people to come across the same section of the installation thus checking the work of past waves. In addition, it allows the crew to be focused on the process of a certain subset of the load-in process. If you have temporary workers, you can save time by only explaining your standards and procedures for the tasks of the day.

For a typical load-in, I use five waves—Hang, Circuit, Color/Templates, Accessories, and Hot Power/Data. Depending on the type of venue that I am working in, I may complete all five waves in one area before moving on to the next. With other venues, I may tackle the entire venue in each wave. For example, if I am working with a counterweight system, it is easier to complete all five waves for one position before moving on to the next one. This way, when I fly the pipe out, it is complete. Conversely, in a catwalk or other grid system, it is easier to complete the hang wave for the entire grid before moving on to any circuiting.

## HANG

For Hang, the technicians will attach all the lights to their positions in the plot. If you are adding positions as part of a rigging project, this should happen in a separate wave prior to the hang wave. Three key standards should be met when hanging lights.

First, all clamps on a position should open toward the same direction. That direction should be determined by which side of the pipe cable will be run along. In a standard proscenium theatre, for example, cable is normally run on the upstage side of the pipe. Therefore, all clamps should open upstage. When working with catwalks, it is generally preferred to keep all clamps open toward the catwalk for ease of working.

Second, all fixtures require a safety cable. The clamps typically used to hang lighting fixtures do not readily show signs of age. Thus, it is possible for them to fail suddenly and without warning. To prevent the lights from falling if this happens, follow the fixture manufacturers' recommended safety cable procedure. In addition to the fixtures themselves, it is good practice to provide a safety cable for any equipment that could fall while manipulating the light at focus such as tophats, barndoors, scrollers, and other accessories.

Finally, all elements of the light that can be tightened should be tightened during Hang. Tightening the clamp is easy to remember, but additionally the clamp's set screw, the fixture's tilt lock, cap screw, and lens adjustment should all be tightened as part of Hang. Tightening these screws and bolts during Hang increases the efficiency of the focus call as all lights will start from the same place.

In many cases, the venue will have a show in it prior to the show you are loading-in. If you are not responsible for that strike, the inventory is likely all in a storage state prior to Hang. In that case, the technicians will focus on moving lights from their storage positions and installing them in their plot positions. However, if you are working in a "changeover" situation where you are leading the strike of the previous show in addition to the load-in of the next, you can add two additional waves prior to Hang—Decolor and Decable.

During the Decolor wave, remove all color, templates, and accessories from the previous plot. Since the installation window is short, I find it best to focus on removing this gear from the venue, but do not worry about putting things into storage until after installation of the new production is complete. For example, when I remove the color, templates, and accessories from lights, I box them up and set them aside in the lighting shop or other safe area until the installation is complete. Once the new plot is ready for focus, I can return to cleaning up from the previous production.

After Decolor, I move on to Decable. During this wave, the technicians unrun the cable from the previous installation and unplug all fixtures. It is also important to remove or reset any temporary power or data infrastructure used in the previous show during this wave. This way the theatre you start the Hang wave with is the same theatre you started with on paper. I call this "zeroing the venue."

If I am transitioning to installation immediately after the completion of strike, I do not strike any fixtures from the previous production unless they are rented or otherwise need to leave the venue. In most circumstances, the production I will be installing will be using the same fixtures as the last one, just in different places. While the new show's fixture positions are certainly different than the previous show's positions, a fixture's storage location is almost always much further away. Putting lights away to take them out again can be inefficient.

For some early-career lighting technicians, there is some confusion as to why returning all the lights to storage is inefficient while striking and coiling all the cable is required. The reason for this is two-fold. First, leaving cable in place creates a significant obstacle for the Hang wave. Hanging lights on pipes with no cable is much easier than hanging lights on pipes that are fully loaded. Second, the portable cable used in theatrical installations is for temporary installation. This requires it to be inspected between uses. Removing it from installation and coiling it gives the Lighting Team an opportunity to inspect the cable to ensure no damage occurred during the previous production.

As part of the Decable and Hang waves, I use a visual tracking method to help the technicians communicate which lights are from the outgoing shows and which lights are for

# THE LOAD-IN

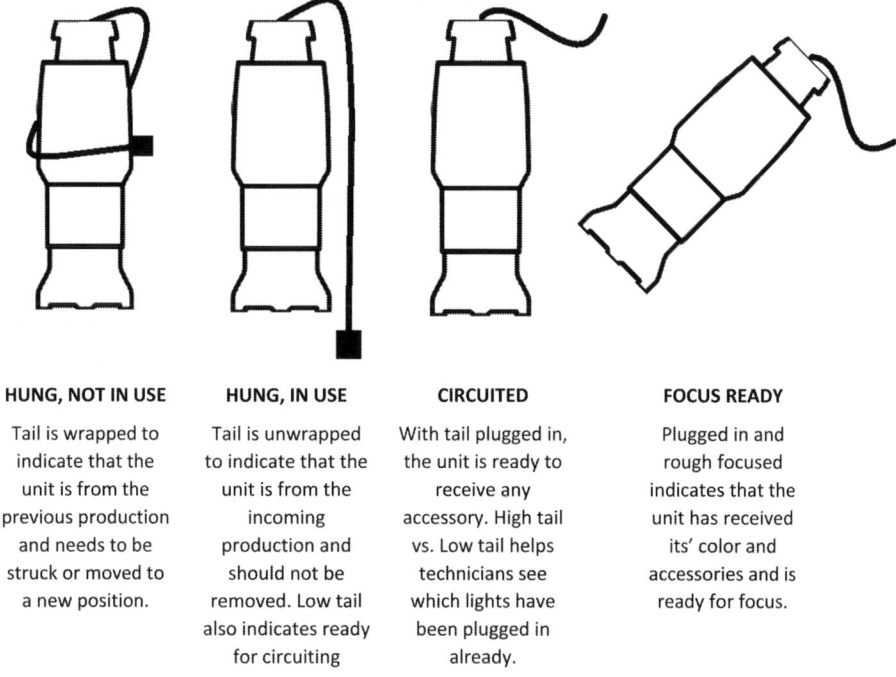

| **HUNG, NOT IN USE** | **HUNG, IN USE** | **CIRCUITED** | **FOCUS READY** |
|---|---|---|---|
| Tail is wrapped to indicate that the unit is from the previous production and needs to be struck or moved to a new position. | Tail is unwrapped to indicate that the unit is from the incoming production and should not be removed. Low tail also indicates ready for circuiting | With tail plugged in, the unit is ready to receive any accessory. High tail vs. Low tail helps technicians see which lights have been plugged in already. | Plugged in and rough focused indicates that the unit has received its' color and accessories and is ready for focus. |

**Figure 9.1** Communicating through position.

the incoming show. Outgoing lights are available to be moved to a new location, incoming lights are installed and ready for the next wave. I call this tracking method, "communicating through position." In this system, "outgoing" fixtures are pointed straight down, with shutters in, and their cable tail wrapped around them. When a light is hung in its' new location, I designate "incoming" by unwrapping the tail so that it hangs straight down and pulling all the shutters. When the Hang wave is finished, any light still in the theatre in the "outgoing" position can be removed and put into storage. "Incoming" lights have their tail hanging down and are clearly ready for circuiting.

# CIRCUITING

During the Circuit wave, I focus on installing cable for dimmed power only. In the same way it is good to separate dimmed power in planning, it is good to separate it here. These dimmed power runs will be the bulk of the cable run for most installations so it best to tackle it first and all at once. If your installation requires multicable to be run to create or expand your circuiting infrastructure, I recommend doing all the multicable runs before doing any additional cable.

When running cable, ensure that all connectors are labeled with at least their circuit number. These labels are important because the cable you run may travel far from its' source. With hundreds of other cables in the theatre, it is important to be able to know quickly which one is which. To aid this process, I recommend that each technician carries a "label kit" consisting of a roll of black gaffer's tape and a fine point silver marker. These supplies will allow them to make quick labels in the field as they work.

In addition to labels, cabling has a couple of standards that want to be adhered to in a professional installation. First, as much as possible cabling wants to be invisible to the audience. To this end, the installation technicians should consider where the audience will be in relationship to the hang position and install the cable on the side of the pipe away from the audience. This should also be the side of the pipe that clamps open on to. If the side of the pipe the clamp opens on to and the side of the pipe away from the audience do not match, it is better to run the cable on the side of the pipe the clamps open on to since it is not ideal for the back of the clamp to become trapped by the cable. In Figure 9.2, you can see that the clamps are opening toward the catwalks in this venue. The cable is run on the side of the pipe with the clamp opening.

In addition to this orientation, the lighting technicians should avoid allowing cable to fall outside of the silhouette of the pipe. Looking again at Figure 9.2, you can see that almost all the cable from the further position is hidden by pipe. I recommend tying the cable to the pipe with tie line at a minimum of 24-inch intervals to help with this.

**Figure 9.2** Cable run on the side of the pipe with the clamp opening.

When tying cable to the pipe use a tight clove hitch to ensure that the cable does not move if tugged. Finish your knot with a bow knot—like you use to tie your shoes—as a safety so that the clove hitch does not loosen. These tie points act as strain reliefs to avoid any pulling on terminals or other connections. Strain relief is important because it prevents separation between terminals or connectors. In addition to stopping the flow of electricity, separation can result in arcing. Over time arcing causes carbon build-up that limits the conductivity of connectors pins or other terminals. Once a connector or terminal has developed this build-up it must be cleaned or replaced.

Furthermore, cable must always follow an architecturally defined path and not travel in open air. Cable ran in open air is much more likely to be snagged on people or scenery during the load-in or production process. Cable that gets snagged can be damaged or in some cases cause trip hazards for technicians working near it. Trip hazards at height can quickly lead to life-threatening situations.

Second, cable needs to be run so that there is no strain on the fixture's circuit tail. This is often referred to as "focus slack" because it is enough cable to allow the fixture the ability to point in any direction during focus. My rule of thumb here is to ensure that the cable servicing the fixture's circuit tail ends within 18 inches of the fixture's clamp. The fixture on the right in Figure 9.3 shows the portable cable terminating almost exactly at the fixture's clamp. This is the ideal situation. While 18 inches is not always sufficient and it is good to reduce that number as much as possible, it covers most circumstances

**Figure 9.3** On the left, each of the four cables going to the fixture have equal focus slack. The fixture on the right demonstrates landing the circuit at the fixture's clamp.

and provides a concrete number for early-career technicians for whom "focus slack" can be a confusing concept.

In many cases, technicians must decide between running cable that is too long or too short. "Focus slack" is an absolute essential, so it is usually necessary for the cable to be too long. There are two schools of thought for dealing with the extra cable—"Dogbones" and Coils. The copper strands inside a portable cable are flexible, but not infinitely so. If you have extra cable in your run, it is best for the cable to coil it into a small coil and then secure it to the pipe near the fixture. Always leave extra cable near the fixture in case the fixture needs to move in tech. Extra cable near the circuit box is largely wasted.

Despite a coil like that being best for the cable, coils have a negative impact on the first standard—keeping cable within the silhouette of the pipe. A common solution to this is the "dogbone." Instead of a tight coil located near the fixture, a much larger coil is made and then compressed against the pipe. The resulting cable—as seen in Figure 9.4—features two low radius turn back points that can be mostly obscured by the pipe. When making "dogbones"—a nickname based on the cable's appearance—it is important to avoid over compression of the turn backs which will break the copper strands inside the cable and result in the cable failure. The amount of compression that the turn back can handle will be related to its' size and number of conductors. Data cable can turn more sharply without damage than multi-cable.

Figure 9.4 "Dogboned" cable.

Additionally, the number of turnbacks should be reduced to a logical minimum. For example, a 12-inch coil with five wraps might become a 30-inch long dogbone with two turn backs. A 30-inch long dogbone is simple to manage. If a dogbone starts to grow to ten feet, it may be more challenging to handle than it is worth. A 12-inch dogbone might have too much cable to hide behind the pipe.

Another common issue arises when a cable is too short, but the next size available is far too long. A technician may wish to use two shorter cables connected to each other. This is not recommended. Using multiple portable cables connected to each other introduces more connectors into a single run. As a matter of practice, runs should be completed with the least number of connections possible to minimize potential points of failure. In this way, a 25-foot cable run for ten feet is more reliable than two five-foot cables run the same distance.

Third, all cable connections need to be strain-relieved on both sides with tie line as shown in Figure 9.5. The only exception to this is the final connector that is attached to the fixture's circuit tail. A fixture's tail should always be left untied with no strain, so that it can easily be unplugged for service. Otherwise, it is important that all connections be securely tied to the pipes they are run along.

Finally, care should be taken to avoid redundant tie line. This occurs when more than one piece of tie line is used to tie up the same point. Redundant tie line is a common problem when different cables are run along the same path at different times. To avoid this problem, tie line should always be untied and retied when running cable along the same

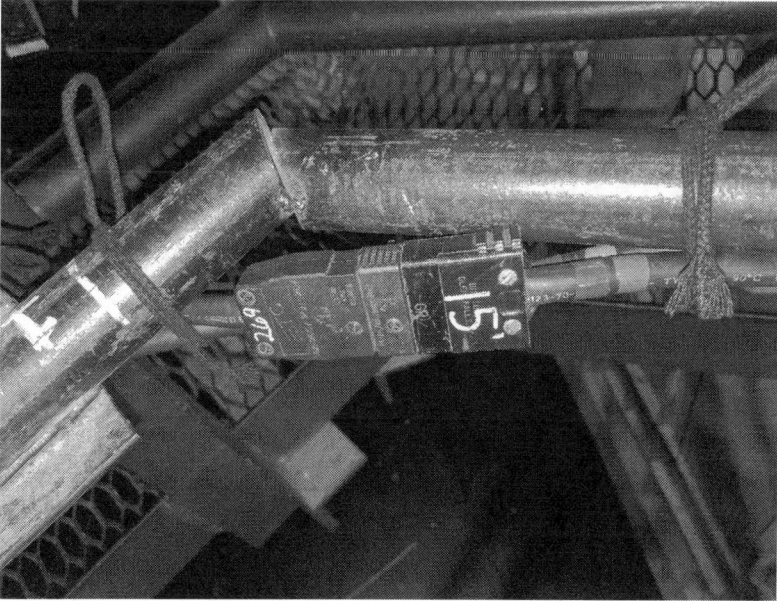

**Figure 9.5** Tying both sides of the connector.

path as already run cable. Redundant tie line can increase the cost of a show by using more tie line than required as well as increase the length of strike due to the extra time it takes to strike the additional tie line.

## OTHER WAVES

After the Circuiting wave comes the Color and Templates wave and the Accessories wave. Depending on the number of accessories, these can be done at the same time. If there are large numbers of accessories, I usually do them as separate waves. When dropping color and templates, you will use the packets you prepared during the Shop Prep phase. Simply go down a position as you go through your packet. If doing accessories at the same time, Lighting Technicians can use the paperwork attached to the packets to know which fixtures require accessories. If doing accessories later, Lightwright has a "limit" feature which would allow the Lighting Supervisor to prepare a separate instrument schedule limited to just accessories.

Once color, templates, and accessories are installed, the load-in moves to the final wave—Hot Power and Data. I recommend doing this wave last in all circumstances because it benefits from the other waves being completed. Using the Hot Power and Data Sheet, technicians will run the planned cables to the appropriate fixtures. Extra care should be taken when labeling these runs. While dimmed circuit labels need relatively little information to be useful in troubleshooting, Hot Power and Data runs benefit from much more. I recommend that each end of the cable be labeled with what it is—Universe *n* or Hot Power Circuit *n* for example—and where it is coming from or where it is going to. For example, in Figure 9.6, you can see a label for a data cable that goes from "x20," short for DMX Output Port Number 20 to "VL@7E#4," which stands for Vari-Lite on the 7th Electric, Unit Number 4.

Otherwise, all the same circuiting standards previously discussed are applicable for Hot Power and Data. "Focus slack" is of particular importance. For example, the fixture on the left in Figure 9.3 has four cables going to it—Data In, Data Out, Power In, Power Out. In this example, all four cables are dressed together with equal focus slack going back to the clamp. This is important because any one cable can impair the focus slack. If the cables jumped to the light from two separate points 18 inches to either side of the fixture's clamp, the cables from one side could prevent the fixture from focusing toward the other side and vice versa.

As an added challenge, most DMX data-grade cable has an outer jacket that cannot withstand the heat generated by the exterior housing of many theatrical fixtures. Thus, these cables must be run in a way that allows for "focus slack" without contacting any fixture housing.

Of course, as mentioned, not all situations benefit from this wave process. When positions require ladder access, for example, it is most efficient to move the ladder as few times

**Figure 9.6** A data cable label: "DMX Port Number 20 to Vari*Lite on the 7th Electric, Unit Number 4"

as possible. Thus, any work that can be accomplished with the ladder in a certain position should be done before moving on. For these environments, I recommend employing the Grounder-Hanger team discussed in Chapter 3. Here, instead of waves, the grounder helps the hanger by prepping the fixtures. The grounder will label the fixtures, select and label any circuiting cable and preload the fixtures with their color, templates, or accessories. With the grounder's aid, the hanger can do the first four waves in one pass before moving the ladder to another section.

## TROUBLESHOOTING

Once all the waves of the load-in are complete, each unit is systematically checked as part of a process called "Ring Out," "Wring Out," or "Troubleshooting." Before beginning this process, the light plot must be set-up in the console. At a minimum this console set-up requires that the starting address of each dimmer and fixture be "soft patched" to a channel. In Chapter 7, we discussed the "hard patch." In the hard patch, each circuit is physically connected to a dimmer with some sort of electrical connection. For the "soft patch" each fixture or dimmer's address is digitally connected to a channel through the console's software. While the Lighting Designer does not care about dimmer and circuit numbers, they do care about channel numbers and will provide them as part of their submission. If using Lightwright or other software, it is possible to import this channel and address information directly to the console without having to engage in manual data entry.

In addition to patching the console, most modern consoles require additional set-up. For example, multiparameter fixtures, like moving lights or LEDs need to be set-up in the console with a fixture profile that tells the console which addresses control which parameters. If you have one of these consoles, it is likely they have a "fixture library" that lets you search and select the appropriate profile. Further, the console is likely to have many system settings, layouts, and other features that should be set up before troubleshooting the plot.

If the console programmer is available, or if you are assigning one of your staff technicians to program the console, it can be beneficial to have them set-up the console instead of you. This way you do not have to worry that later changes that they need to make to the patch, profiles, or other set-up options will cause something to unexpectedly stop working. You want to troubleshoot a plot that is in its' finished "pre-focus" state, so that you know that it will work properly during the focus call.

Also, because most consoles have offline editors, it is completely possible—and advisable—to set-up the console during your Shop Prep time. If you can set-up the console early, at load-in you can simply upload your file—or a file from the programmer—and get started with troubleshooting.

When troubleshooting, you want to make two full system passes. On the first pass, you will check each unit in the order it appears on its' position, following along in your instrument schedule or on the plot. During this check, you will ensure that the light comes on, it is in the right spot, and appears to have the correct accessories. It is good to do this first check in position order as it helps you determine if the correct light is coming on due to its relationship with other lights being checked on that position.

On the second pass, you will print from Lightwright a new document called a "Channel Hook-up." This will list the units by channel instead of by location. Prior to this point, your work has been primarily concerned with getting lights to certain places, so the Instrument Schedule was the ideal format for your information. At this point, you will start to refer to units in the same way that the Lighting Designer does—by channel. Given that, your second pass will allow you to check lights by channel. Normally, troubleshooting by channel will group lights with similar purposes in sequence. This pass should allow you to clearly see any color or template mistakes that were close enough to be missed on the first pass.

Finally, the Lighting Supervisor checks each parameter of any multiparameter device in the rig to ensure that the correct fixture responds as expected. Multiparameter devices are any fixture or accessory that does something other than just light up. It is important that all parameters are checked. You do not know which parameters will be important for the production, so it is best to assume that if a certain light was specified, the Lighting Designer wants access to all parameters. That means that you need to ensure that all parameters are functioning. For example, a moving light that moves, but does not change color would likely cause quite a disappointment. Checking these fixtures now should give you time to fix or replace them prior to focus.

When troubleshooting the rig, remember that the goal is for all load-in tasks to be complete and functional prior to focus. If you do not have a blank to do list when focus starts, the tech process will be unnecessarily challenging and that will impact your ability to establish peership and trust with the Lighting Designer during the final steps of the process. Troubleshooting is your final check to ensure that the rig is ready to hand back to the Lighting Designer.

CHAPTER 10

# Focus

## PREPARING FOR THE BIG GAME

Focus is the culmination of all the engineering work you have done to prepare and install the plot. With the equipment fully functional and your to do list clear, it is time to hand the baton back to the Lighting Designer. Focus represents the first time that the Lighting Design team and the Lighting Engineering team are working together in the same room. For that reason, it is good to break down a little bit of who is in there and what their role is.

The Lighting Designer takes centerstage during focus both figuratively and literally. They will be onstage directing the team of lighting technicians as to where the lights will be pointed. Focus is the Lighting Designer's show and as much as possible the Lighting Supervisor and the team of technicians are merely facilitating it. You want to give the designer space to work. However, there is usually a time crunch at focus, so you want to be mindful of anything you can do to help keep the designer moving forward.

The Lighting Technicians are doing the physical focus work and manipulating each fixture based on the designer's direction. If ladders or hydraulic lifts are being used, technicians should be paired up so that there is a focuser and a grounder. Working in pairs helps keep ladders or lifts moving. Logistically, it may not be possible for each focuser to have a dedicated grounder. If that is the case, try to maintain as close of an even focuser-to-grounder ratio as possible. For example, if you have only two technicians, it is generally better to have one focuser and one grounder rather than two focusers and no grounders. If you have three technicians you can have two focusers and one grounder, but it will be important that the one grounder be aware that they are working with both focusers.

When determining the number of technicians needed for focus, in addition to any grounders, remember that special projects will arise during the call that will require you to pull people away—a lamp might blow and need to be changed or a fixture may need to be moved to a new location because its' angle does not work. In an ideal world, you should plan so that enough technicians remain available to work with the designer even when some must peel off to handle these issues. For example, consider a call with four focusing technicians. This allows two different side projects to be managed while still maintaining two technicians to work with the designer.

Ultimately, your labor budget will determine how many technicians you can have available for focus. Therefore, it is important to consider the staffing of the focus call during your initial Price Out. If the plot you receive is small, you may be able to make do with less focusers. In this case, putting the designer on a break while the technicians work on a special project may not be problematic. However, if the plot is large or otherwise complex, pausing the designer's workflow to fix a problem or move a light could cause the call to run over its' scheduled time. This additional time could result in increased labor expenses and a delay of other scheduled on-stage work.

To help in evaluating my labor needs, I have taken to keeping focus speed statistics. I record the total amount of time the focus took and how many lights were focused in that time frame to get a "minutes per fixture" rate. In general, I have found that one and a half minutes per fixture is a good average speed. Tracking these statistics allows me to know which designers work faster or slower than others as well as which venues or technician counts result in faster or slower speeds. This way, I can use past data to anticipate how long a focus will take given the venue, designer, and technician count.

For example, if my crew is normally four people and we have an average speed of one and a half minutes per fixture, a three-hundred-unit plot will take approximately seven and a half hours to focus. To preserve some contingency in the event of a large number of special projects, I add an hour to my estimate. When planning the schedule with other departments, I will make sure they are aware that this focus call should take eight and a half hours. If that amount of time is unavailable, I will need to determine if more focusers will decrease the duration of focus or work with the designer to scale down the plot.

## CALLING THE FOCUS

During the focus call, the Lighting Supervisor normally takes on one of three roles. Which role they take is usually dependent on the size of the Lighting Design team. In many cases, the Lighting Designer works by themselves. If that is the case, the Lighting Supervisor will "call focus" and keep the crew moving efficiently. If the Lighting Designer uses an assistant or associate, it is often preferable for this person to be the one calling focus. An assistant or associate often has an existing relationship with the designer, which allows them to anticipate the needs of the designer and keep them moving forward.

If working with an Assistant or Associate Lighting Designer, always ask them if they would like to call focus. It is good relationship building to give them the option rather than assume that they will. In some cases, they may prefer that the Lighting Supervisor calls focus so that they can work with the designer more directly to make or read focus charts—documentation of the focuses of each light—or perform other assistance.

If the Lighting Supervisor is not calling the focus, they are either working as a lighting technician—increasing the number of available focusers or grounders—or working as a floating technician to handle special projects so that the focusing technicians are not pulled off of their primary task.

If calling focus, there are certain best practices. First, you will need a clear and reliable way to track where technicians are, which light is currently being focused, and which lights are complete. For this, I normally use a combination of a printed light plot and the digital Lightwright File. On the light plot, I highlight any light that I send a technician to in one color—as shown in Figure 10.1. Once those lights are focused, I highlight them again in a different color. This gives me a quick graphical representation of where we are.

In Lightwright, I use the "Focus Status" column and one of the user columns to track focus. The Focus Status column is a toggle between three states "Focused," "Partially Focused," and "Not Focused." As you can see in Figure 10.2, the red circle with an "F"

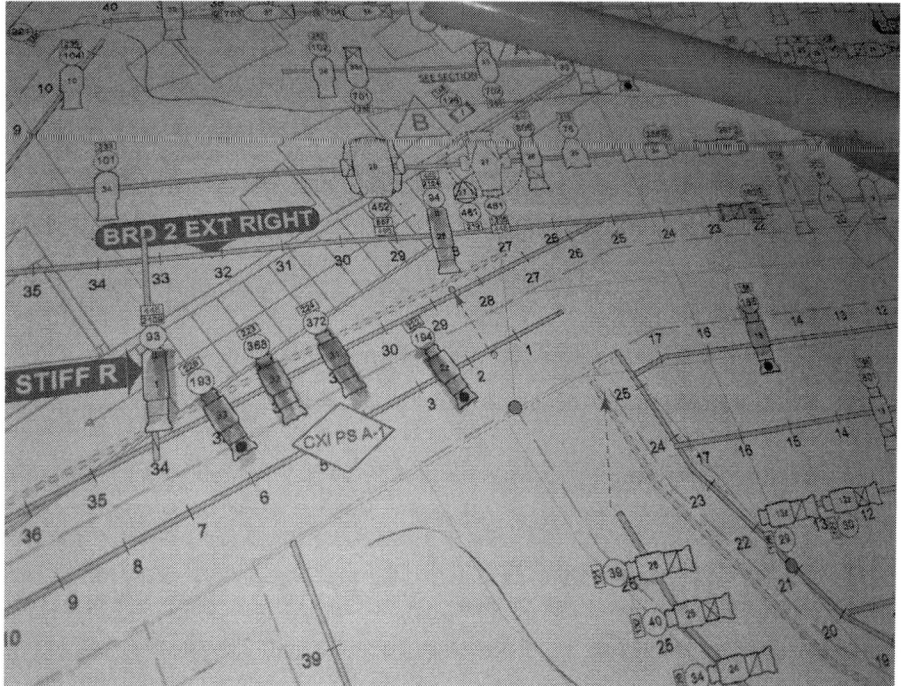

**Figure 10.1** Highlighting the plot during focus.

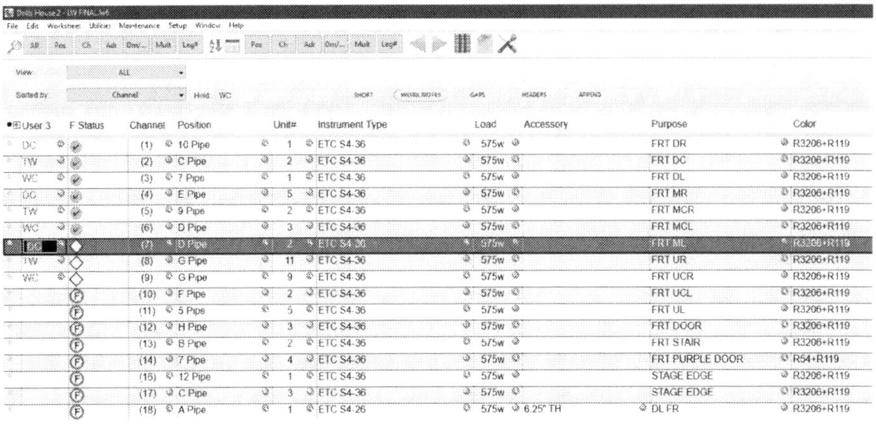

**Figure 10.2** Focus status in Lightwright.

means "Not Focused," the yellow diamond means "Partially Focused," and the green checkmark means "Focused."

When I send a technician to a light, I mark it as "Partially Focused" to note that someone is currently working on it. In the user column, I record the initials of the person I sent, so that I remember who is on which light. Finally, when the light is focused, I change the status to "Focused."

To reap the maximum benefit from the Focus Status column, it is important to setup each fixture's "focus ability" in the Instrument Type Maintenance. I normally do this as part of my initial Plot Clean up as discussed in Chapter 6. By default, all fixtures types are set to be "focusable." The focus ability column—as you can see in Figure 6.3—is represented by the "Not Focused" symbol. Each entry with this symbol is "focusable." If you have an entry that does not get focused—like a color scroller or a moving light—you can click that symbol and it will go away. These entries are considered "not focusable."

On your Lightwright worksheet, you have an option to filter your display by using the dropdown menu called "view." If you set your view to "Needs Focusing," only fixture types that have been set to "focusable" in the settings will appear.

When I am calling a focus, I bookmark this view using Lightwright's bookmark feature. As I mark lights as "Focused," they no longer meet the criteria of the "Needs Focusing" filter, but do not immediately disappear. To rerun my filter, I simply recall my bookmark. Now any completed lights are hidden, and I am left with a worksheet that shows only the lights that still need to be focused. Conveniently, Lightwright also provides you with a count of fixtures in your worksheet. Since I have filtered to only see lights that need to be focused, refreshing my bookmark gives me a quick count of remaining focusable fixtures.

I can use this count to estimate whether or not we will complete our work on time. If I see us falling behind, I can send a quick message to whoever is planning to be on stage after us to give them a head's up.

Second, you need to be prepared to record any changes or take other notes that might occur during the focus call. Focus moves fast and if you miss a change it could be left out of the final paperwork. Always be aware of the conversation that is happening on stage. This is another reason that I like having the digital Lightwright open and ready during the call. For example, if a designer decides to change the color of a light, I can record it immediately. Now, next time I print paperwork it is already correct. If I do not record that color change and we need to replace the color in the middle of the run, we might very well accidentally restore the color to old choice.

Sometimes, of course, things move too fast to edit the Lightwright workbook without making a mistake. For those situations, I always keep paper and a pencil next to me. If there is a change that cannot be done in the moment or a paperwork adjustment that needs to wait to ensure accuracy, I will jot it down to handle later. Sometimes, I will make these notes directly on the light plot because it allows me to use shorthand. For example, if a light moves to a new location, it is easier to circle it and draw an arrow to its' new location than to try to describe the move in words. Ultimately, the plot used for focus should be discarded and reprinted prior to tech, so is no reason to not draw on it.

Regardless of how your record your notes, be sure to take a moment at the end of focus to compile all of them so that they will make sense to you later. In the heat of the moment you might circle a unit and write, "TH." That is great for the moment, but as you prepare your to do list for the next work day, be sure to write it out so that you or anyone else can easily understand—"Add Tophat to Channel $n$." Like your end-of-day reconciliations during the installation process, this is also a great time to make sure that any changes are clearly represented in all of the paperwork.

Finally, have a plan for the day. In fact, have ten plans. Discuss your plan with the Lighting Design team before the day starts. When the focus gets going, you, as "caller" will be dictating which light the designer goes to. Make sure that the plan you have decided on works for everyone. Also, be prepared for that plan to fall apart when some of your technicians must work on repairs or other side projects. Really try to understand how the plot works in advance of focus and look for areas that will slow things down from either a design perspective or a logistical perspective. Try to see if there are ways to mitigate those delays.

When I develop the focus plan, the first thing that I weigh is designer ease versus technician ease. For example, if I decide to "focus by position," I can put technicians on a catwalk focusing each light in the order they occur on the pipe. That is fast and easy for the technicians. However, the designer is now bouncing wildly from front lights, to specials, to side lights, and so on. So, even though the technicians can work quickly, the designer may slow down as their brain needs time to process before each fixture thus hurting overall efficiency.

Alternatively, I can "focus by system." Here, I use the designer's purposes to focus lights with the same or similar descriptions in groups. For example, I can focus all "front light from left" then all "front light from right" and so on. Most designers group their systems numerically, so when focusing by system, you can simply start with Channel 1 and work your way down the Channel Hook-up. This allows the designer to know which light will be next and where it goes before you announce its' channel. Of course, the technicians are now physically running around the room.

Often there is a balance. Lights that are challenging for technicians to access are best done "by position," whereas lights that are easy for technicians to access should be done "by system." When discussing your plan with the designer, be sure to let them know whether you are focusing by position or by system or with some combination.

The second thing I consider are access issues. Sometimes, the focus of one light makes it harder to focus another light. For example, when working on a boom or other vertical position, it is often better to start at the top and work down. This is because as the technician climbs down the ladder, they have an opportunity to bump any the lights as they pass them. If they only pass unfocused lights, this concern is largely mitigated.

Other things could also present access issues. A ladder, for instance, has a defined footprint. It is hard for two ladders to exist in close proximity. Additionally, if one focuser is trying to focus a light where another focuser's ladder is, it may be difficult for the Lighting Designer to see what is happening. Not all these situations can be avoided, of course, but if you consider them in your planning you will be able to predict what their impact will be.

Finally, it is important to consider the purpose of the lights. Area lights will always focus faster than specials. If the light is focused on a scenic element, the designer may be able to instruct the focuser and move on while the focuser works independently. Some lights require a specific scenic configuration while others may be hindered by one. Because of this, it may not be logical to simply start with Channel 1 and work down the instrument schedule.

For example, a plot might have large number of cyc lights and a handful of specials. You know the specials will slow the process down, but also that the cyc lights can be done without much designer input. If you plan to focus the cyc lights and the specials at the same time you can work on multiple lights simultaneously thus lessening the impact of the slow down caused by focusing specials.

I always note lights that I think can be focused without direct designer input. If we come to one, I will ask, "can they work on this for a bit while we move on?" or "can you show them how to do this system and give them some time to work on it?" The designer will understand what you are after right away. Sometimes they will disagree and want to be part of that focus, but there is rarely any reason not to make those sorts of suggestions.

## TECHNICIAN FOCUS STANDARDS

Because the pace of focus must be so quick, it is important that the technicians understand what is expected of them and work to develop efficient focus technique. Again, each light should only take one and a half minutes. This is a goal for the designer as well as the technician. Of course, the technician cannot control the pace of the designer so the best they can do is ensure that the designer is never waiting for them. To work toward that expectation, I encourage technicians to follow a defined sequence for each light they focus:

- Hear the assignment and confirm.
- Get to the light quickly and safely.
- Pull color and go sharp to shutter or gobo.
- Loosen and tighten.
- Watch and listen for similar focuses.
- Center designer in hot spot.
- Watch and listen for instruction.
- Lock it and confirm.
- Set shutters and run lens tube.
- Restore color and confirm completion.

To start with, the person calling focus—the "caller"—will provide the technicians—the "focusers"—with direction. Focusers should listen for their assignments and keep conversation to a minimum so that others can hear theirs. When they hear the assignment, they should respond with a confirmation so that the caller knows that it was heard. A simple, "got it," "on my way," or "thank you" is usually sufficient. When the caller assigns the light, they should help the technician find it by bringing it up to a low level. This level needs to be low enough as to not interfere with any lights being focused, but bright enough so that the focuser can see the beam that it casts. I recommend between 20% and 30% depending on the size of your venue and the efficiency of your fixtures.

The focuser should get to their light as quickly as possible without creating any dangerous situations. Moving too quickly has the potential to result in injury which will slow the process down even more. I usually encourage technicians to consider the difference between the way tourists walk down the street and the way life-long citizens do. In this case, be a citizen.

When they arrive at their light, the focuser should immediately pull the color. This is done for two reasons. First, saturated colors make the light harder to see. Both the focuser and the designer will benefit from a brighter fixture when setting the focus. Second, most designers use some sort of diffusion gel to make the edges of the light fuzzy. This is great for the show but makes it difficult to see the edges of shutters and beams on ellipsoidals when you are focusing.

With the color pulled, the focuser should point the light roughly where they think it will go. The goal here is to emulate the distance of the final focus so that they can sharpen the edge. If they point the light at a spot that is much closer or further than the intended throw of the unit, their sharpening will be inaccurate, and they will have to do it again.

When sharpening the fixture, they can choose to sharpen to the edge of the beam, the shutter, an iris, or a template. Since none of these parts of the light exist in the same plane, sharpening to one will prevent another from being sharp. If the designer does not share a preference at the top of the call, always default to a standard priority—template, iris, shutter, then beam. The first item that their light has is the item they should sharpen to. Make sure that the focus is as sharp as possible. If they do not sharpen enough, they will have to do so on the designer's time and that will make the process take longer. Much like prep, if you can do it in advance, you should.

Not all lights, of course, can be made sharp. If there are soft focus wash lights in the light plot, the caller should clarify with the Lighting Designer in advance if those lights need their color pulled for focus so that all focusers can know. In many cases, the color does not need to be pulled on these lights and skipping that step when unnecessary saves time.

After the fixture is sharpened, the focuser should check all the common knobs and bolts on the unit. Some knobs or bolts will want to be completely tight and secure. Others will want to be loose enough so that the fixture can be easily adjusted during focus, but not so loose that tightening them later takes more than one half turn.

Bolts that should be completely tight include the lens tube knob and the clamp's set screw—also called a "Jesus Bolt" among other colorful names. Bolts and knobs that should be partially loose are the tilt lock handle and the yoke bolt. The yoke bolt should be loose enough to allow for a smooth pan over the fixture's full range while not causing the fixture to rock. Likewise, the tilt lock handle should be loose enough to allow for full range of motion while not wiggling in its' slot. In an ideal world these and all other potentially loose elements of the fixtures were checked and tightened during hang so setting them to their focus status should be quick and easy.

With these bolts and knobs set, the focuser waits until it is their turn to work with the designer. While they are waiting, it is important to remain attentive to what is happening. Most of the lights in the rig will be designed to work as a system. Usually a focuser can predict where their light will fall and how the shutters will be set by watching the designer work with others. While the focuser should never take the liberty of completing a focus before the designer gets to them, roughing in the position and then being prepared to understand any shutter shorthand is useful in ensuring that the pace of focus is maximized.

When the focuser's turn to work with the designer arrives, the caller will bring the light to full and announce its' channel number and purpose. This is a signal to everyone which light is now up for focus. The focuser will respond by centering the designer's head in the beam and respond to any directions they receive. It is important that the focuser situate themselves such that they can manipulate the light as well as see the designer. The designer will frequently use some combination of vocal direction and hand gestures. If the focuser cannot see the designer, they will not be able to properly connect with them.

During this process, if any issue arises that prevents the focuser from completing the task at hand, they should ask that they be skipped rather than allow the designer to wait on them. Any issue they have is normally better resolved while someone else is working with the designer. In an ideal situation, the technician who is focusing will try to correct for any potential issue in advance. For example, if the light is only able to turn one direction because it is colliding with a structural member of the venue, the technician may be able to slightly adjust the position of the light to allow full rotation without significantly impacting the angle of the shot. Dealing with these issues prior to working with the designer has a significant impact on the overall pace of focus.

When the position of the light is set to the designer's liking they will normally say "lock" or make gesture that means it. It is important that the focuser fully lock the position of the fixture prior to making any shutter or lens tube adjustments as either adjustment will cause the position of the light to shift if it is not locked. To lock the fixture, tighten the yoke bolt and tilt lock all the way. A wrench is required for one or both adjustments. With all bolts and knobs now tight, the focuser should announce, "locked." This tells the designer that they can move on to shutters and lens tube adjustments.

After the focuser makes the designer's requested shutter and lens tube adjustments, they will restore the color to the fixture and confirm that the light is now complete. This can be done by announcing "done" or "thank you." With the light complete, the caller will turn it off and provide the focuser with a new assignment thus restarting the process. The focuser needs to remember to be immediately on the lookout for their next assignment as they may need to travel to a different part of the theatre and they will want to get the next light ready before it is brought to full.

CHAPTER 11

# Tech, Performance, and Strike

## TAKE A STEP BACK

The first thing I do when I finish focus is breathe. Inevitably, focus will have generated notes and changes. Certainly, the work is not yet complete, but when focus ends—as long as it went well—you have done most of your job. Sometime ago you received a plot submission and after careful conversations with the Lighting Designer, you were tasked with making that drawing into a safe and functional reality. With focus behind you, you can sit in the venue and admire that work.

Now you hand the rig back to the Lighting Designer. This can be unnerving. It is much like a shipbuilder launching their ship into the sea or a software developer sending a new application into beta. Your work is about to be put to the test. Does it work as it intended? Does it need repair? Does it need changes?

I often reflect on this coming moment as I plan and install the plot. I will ask myself can this rig be used to its' full capacity and still not break? Am I giving the run crew a solid installation or is it one that is riddled with failure points?

As an example, there was a production I worked on that had a revolve. The revolve carried a two-story house with many set electrics and practicals. It was important that the revolve turns be counted and then reversed each day because the cable that provided power to it would twist with each turn. Too many turns in one direction would result in damage to the wire. This was not a great plan. The system had a known failure point that the crew had to avoid. The next time I had a similar challenge, I used a device called a slip ring that I purchased from a "do-it-yourself" wind turbine company. This allowed unlimited rotations while still guaranteeing contact. Always remember that you are building something that will be used by other people and work to minimize potential failure points.

With everything ready to go, next on the schedule are technical rehearsals—or "tech" as it is normally called. What is the role of the Lighting Supervisor during tech? It depends heavily on the show, but in a perfect world they would do nothing. Everything they built would work perfectly, everything the Lighting Designer planned for is exactly what the show needed, and all the paperwork is updated and finalized. Unfortunately, this is not the reality most of the time.

## NOTES CALLS

Starting as soon as the Lighting Designer's plane lands or their foot crosses the venue's threshold, you will start to get notes. If you followed my recommendations, you would have your blank notepad ready to receive them. Your installation is complete and you are ready to start changing it. Tracking notes is an important part of staying organized during the tech process and there are a few different ways to do it.

The easiest way to take notes during the tech period is on a notepad with a pencil. It seems obvious to some, but despite the multitude of other useful notetaking options, nothing fully replaces a notepad and pencil. I always have one even if I am using a digital note taking system. There will almost certainly be a moment where you need to take a note and your computer or tablet is powered down or out of batteries. Not taking the note is not an option. Do yourself a favor and always have notepad. Mine lives on a clipboard so that I can keep paperwork underneath it if I need to.

The two most important things about taking notes is to categorize and to ask clarifying questions. For categories, I leave space along the left margin to write a category letter for each note. If I need to "Change Channel 7 and 17's color to Rosco 53," as you can see with the second note in Figure 11.1, I will categorize it "W/PW." The "W" stands for "work." "Work Notes" are ones that I or a technician will have to do during a period when rehearsal is not happening. These notes require the space be made safe for overhead work and likely be scheduled with Stage Management. The "PW" stands for "Paperwork Notes." This is a reminder that I cannot cross this item off until it has been recorded in the show's paperwork. Other categories that I use are "F" for "Focus Notes"—meaning

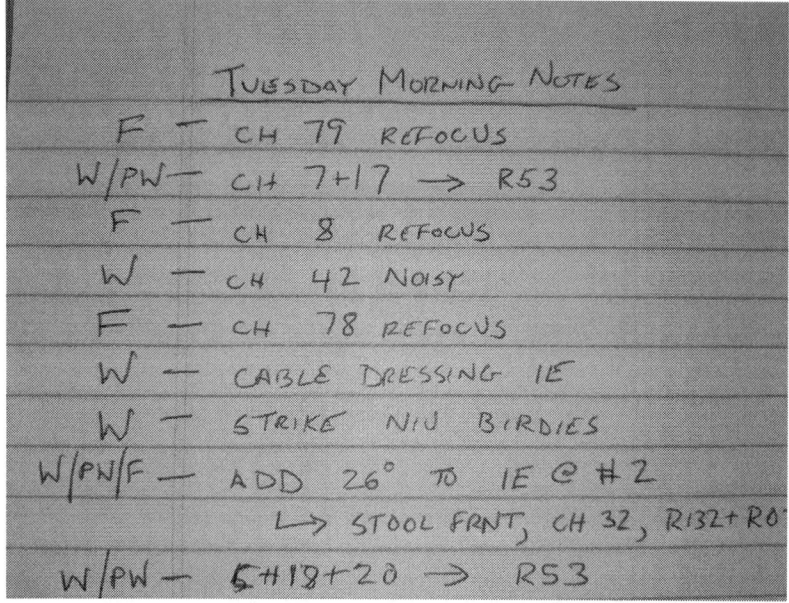

**Figure 11.1** Taking notes on a notepad.

I will need to schedule dark time with Stage Management and ensure that the Lighting Designer is available when I want to do the note and "TD" for "I need to discuss this note with the Technical Director"—usually meaning that the Lighting Designer has a note for scenery and I need to help facilitate the conversation. Invariably, certain circumstances require additional categories, but these cover most situations.

Second, since most of the notes that go on your notepad are not ones that you came up with—although there will likely be some of those too—it is important to ask questions so that you fully understand them. Most of the time you will receive notes from the Lighting Designer, but they will not be there when you execute the work. Ask the designer for additional information as they give the note so that you know what they meant when the time comes.

For example, when a Lighting Designer wants to add a light, they will normally point in the air somewhere vague and say something like, "let's add a 36-degree over there." I like to keep a copy of the light plot on my clipboard—updated and reprinted post-focus—to take out at times like this. I point to the drawing to confirm the location. Normally, I am glad I do because half the time I get it wrong. With the position confirmed, I draw a quick symbol—like an arrow—on the plot so that I remember then make the "Add 36-degree" note on my notepad. Before the designer moves on, I am also sure to ask,

"Color? Channel? Purpose?" so that I have an entry for all three on my sheet. You can see a note like this as the second to last one in Figure 11.1. With this recorded, I have enough information to both get the light ready for focus as well as update the paperwork.

If the Lighting Designer has an assistant or associate, they will often be the intermediary between you and the Lighting Designer for notes. This helps immensely, as the Lighting Designer can give their assistant notes in real time and you can step out of the room for a well-deserved break without worrying about missing something important. I normally take my notes in the same way even if there is an assistant. I just get the notes from the assistant instead. Any good assistant will be as knowledgeable about the note as the designer would be.

Additionally, when there is an assistant, they are normally responsible for updating the paperwork. Again, this can be very helpful, but also remember that some of the paperwork you made was to track the rig's infrastructure. That paperwork is always yours to update. The assistant will take care of the Lightwright file and the drafting. They will even generate all sorts of additional paperwork to track cues, spotlights, and moving lights, but your Hot Power and Data Sheet, Mult Sheet, or other such documentation is all yours.

If there is an assistant on the project or the show is particularly complex, using some form of digital notetaking is useful. An online spreadsheet program—like the Google Sheets document shown in Figure 11.2—is great for this. The assistant can set-up the notes list and the appropriate category designations and give themselves a short form to fill out to ensure they get all the details. Then, all the assistant needs to do is send you a share link and you can watch the notes come in as they type them. When you complete the notes, you can mark them as done.

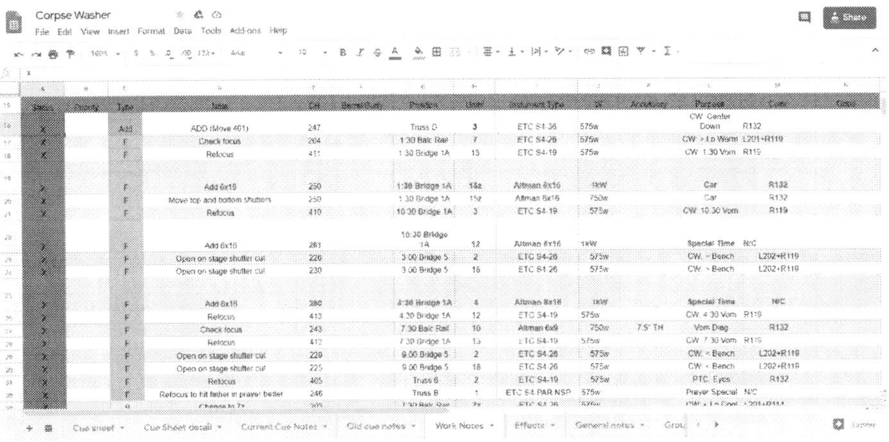

**Figure 11.2** Taking notes on Google Sheets.

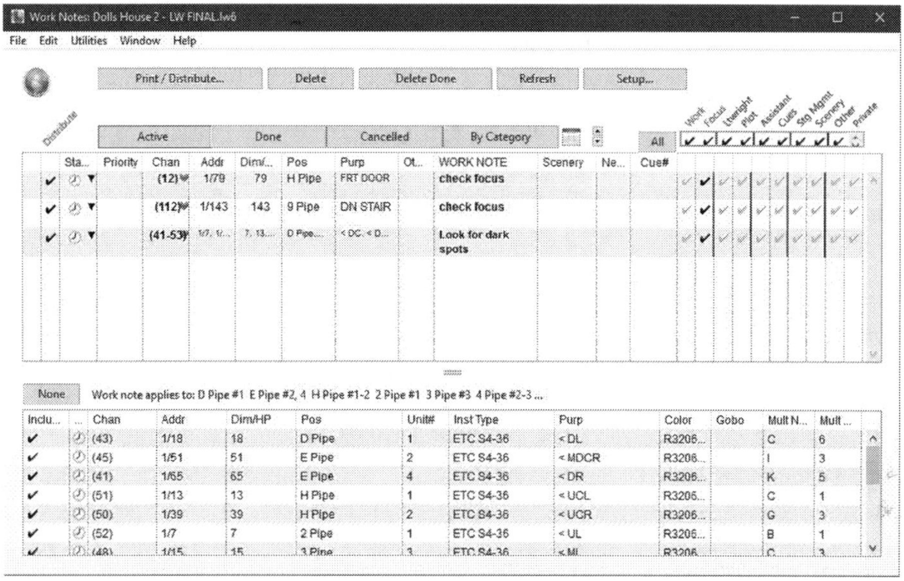

**Figure 11.3** Lightwright's worknotes.

Finally, Lightwright has built-in note taking ability. There is a bit of a learning curve with the Lightwright notes, but if you can get the hang of them, they offer many great features. For example, each note in Lightwright can be tied to a fixture. This allows you to have all the information about a fixture connected to the note. If you want to make a note about adding a new fixture, this forces you to add the fixture first which will help you remember to ask about its' details. Additionally, Lightwright offers a host of options to print or e-mail notes to collaborators.

Regardless of how you record the notes, running the notes calls themselves will work like a mini-installation but with increased time pressure. Each call will require that you make a plan that is prioritized for efficiency and coordinated with the other department's work. As you look through your notes list, determine which notes must be completed before the start of the next rehearsal and which can wait. Think about which notes can be done at the same time. Look at which notes can be done outside the room—like paperwork updates and practical repairs—and save them for when you cannot access the stage. Some notes calls will allow you to finish the entire list, but many times notes will roll over into the next call. It is normal not to finish the list, but you must make sure that you prioritize the things that need to be completed before the next rehearsal and continue to track the things that are still pending.

Most notes calls will end in focus or cueing time. For this, the Lighting Designer will need a dark or nearly dark stage. Make sure to coordinate that need with Stage

Management and other department managers. Bear in mind that Stage Management has a responsibility to ensure the stage is clean and safe for actors prior to the start of rehearsal. Work with them to preserve time for them to do that cleaning in full light.

As Lighting Supervisor, you will set the lighting team's call times—including that of the designer. You will need to determine when you need technicians for notes, what time the Lighting Designer is to arrive, and when to call your console programmer or other crew. There is no reason to call people earlier than they need to be there, but there is also no reason to be afraid to call people early. For example, it is common for early-career Lighting Supervisors to delay the Lighting Designer's call time as late as possible. While this is certainly appreciated by the designer, it is a problem if it results in hindering Stage Management's ability to prepare the stage or if the designer is unable to complete all their notes. In many cases, the Lighting Designer will trust that you know how long things will take and what your crew needs. Be sure to think through each call before assigning times.

## EMPOWERING THE CREW

In addition to supervising the installation crew, the Lighting Supervisor also hires and manages the tech and running crew. These technicians include the Console Programmer, Light Board Operator, Follow Spot Operator, and Deck Electrician. Not every show requires all those roles and some shows will need others. It is also customary for some roles—like Console Programmer and Light Board Operator—to be combined when it is economically necessary to do so. The Lighting Supervisor needs to work with the Director of Production and the Lighting Designer to plan for and balance the needs of the show with the fiscal capacity of the producer.

The crew the Lighting Supervisor hires will need to be well-qualified in their areas and the Lighting Supervisor needs to trust them to be effective. During the tech process, you will need to focus on managing notes and then in the later run of the show you will likely need to move on to your next project. Having technicians who can work on the show independently is crucial toward ensuring that you will have the time you need to fulfill your other tasks.

Unfortunately, some budgetary restraints will cause certain producers to be unable to hire well-qualified individuals to serve on a show's running crew. In this situation, the Lighting Supervisor still must put trust in these individuals. When possible, I recommend using time outside the scheduled work to help these technicians expand their skills and to provide clear feedback on their performances. For example, many small theatres use interns or apprentices in place of staff Console Programmers or Operators. Normally, these early-career technicians have not had the level of exposure to the console desired for such a role. To compensate, you can provide them with dedicated console training time that allows them to gain hands-on hours outside of tech and become better prepared.

Regardless of the level of experience a technician may have, their success is always contingent on your providing them with the tools they need to do their job effectively. It is important that the technicians who will be running the show fully understand and can take ownership over the rig. This can be achieved through detailed documentation, in-person walk throughs, and checklists—as in Figure 11.4. It is important that the

## BOARD OPERATOR CHECKLIST

**PRESHOW BEFORE HALF HOUR**
- ☐ Disable Fire Alarms
- ☐ Turn on Eos Console
- ☐ Load Show File (if necessary)
- ☐ Turn on Hot Power
  - ○ Console Macro
  - ○ Rigging Room CC Modules: 309, 310, 414
- ☐ Uninhibit Strobes and Fog
- ☐ Drummer TV Power Sub Up
- ☐ Warm halogen lamps at 05%
- ☐ Switch on Hazer (Bridge 6)
- ☐ Switch on Foggers and press "DMX" button (OSR and Trap Room)
- ☐ Change Trap PARs to RED Gel Frame
- ☐ Switch on Drummer TV
- ☐ Install Batteries:
  - ○ Neverland
  - ○ Wasp
  - ○ Piano
  - ○ Lantern (one 9v per week)
- ☐ "Go to Cue 0" on iPad
- ☐ Run Channel Check
  - ○ Look for lamp functionality
  - ○ Look for focus and shutters
  - ○ Look for color integrity
  - ○ WARN ON STROBES
- ☐ Test Moving Light Parameters
- ☐ Test Scroller, Rotators, DMX Irises, I-Cues
- ☐ Test Foggers and Hazer (Remember to Warn)
- ☐ Ensure Spot Operators look at their lights:
  - ○ Blue Light
  - ○ Movement
  - ○ Flags and Iris
- ☐ Take Control of House Lights
  - ○ Run up House Submaster on the console
  - ○ Press "All Out" on unison
  - ○ Press "Manual" on unison to lock out
- ☐ Turn on Aisle Lights
- ☐ Turn on Infrared

**AT HALF HOUR**
- ☐ Blackout Check
  - ○ In coordination with SM
  - ○ Turn off All Work Lights
    - ▪ HIDs
    - ▪ CFL Master
    - ▪ 3 CFL Switches in Booth
    - ▪ Upstage Works by Roll Door
  - ○ Run down House Light Submaster
  - ○ Ensure that full blackout is achieved
    - ▪ Look for light leaks
    - ▪ Look for open grid door
    - ▪ Look for open garage door
- ☐ Put the console in cue
- ☐ Run down all Additive Submasters
- ☐ Run up all Inhibitive Submasters

**INTERMISSION**
- ☐ Change Trap PARs to BLUE Gel Frame

**POST SHOW (ONLY WHEN AUDIENCE IS CLEAR)**
- ☐ Restore House Lights to Panel Control
  - ○ Toggle to "Day" state.
  - ○ Add Grid Works (3 Switches in Booth)
  - ○ Switch on CFL Master
  - ○ Add HID Works (if needed)
  - ○ Press "Manual" to unlock panel
- ☐ "Go to Cue Out" on Console
- ☐ Turn off Hot Power
  - ○ Console Macro
  - ○ Rigging Room CC Modules: 309, 310, 414
- ☐ Turn off EOS Console
- ☐ Collect and Charge Batteries:
  - ○ Neverland
  - ○ Wasp
  - ○ Piano
  - ○ Lantern (one 9v per week)
- ☐ Turn off Infrared
- ☐ Turn off Aisle Lights

ALWAYS: REMEMBER TO KEEP THE BOOTH CLEAN AND STOCKED AND GRID IMMACULATE! | 1/24/2016 10:23:08 AM

**Figure 11.4** A typical daily checklist for a Light Board Operator who also has show maintenance responsibilities.

technician feel like an expert on the show even if they are still working on being an expert in their career.

The documentation that I normally provide to a running technician is as follows. I try to provide all paperwork as a hard copy and as a digital copy.

- Instrument Schedule
- Channel Hook-up
- Data and Hot Power Sheet
- Mult Sheet
- Daily Checklist

Most of this paperwork will look familiar as you have been crafting it throughout your prep process and using it throughout your installation, focus, and technical rehearsals. Handing it off to the programing or running technician can feel like passing the baton. In a way, that is exactly what you are doing. You are saying, "Here is the rig I built, take care of it." In most cases you will remain available should any issues arise but shifting the day-to-day management of the rig to the running team is an important transition. Be confident that you built a good rig and confident that your technician will mind it well.

With the show running, work with the running technician to develop some sort of regular reporting system. Much like a Stage Manager's performance reports, these reports can help you know of any issues that might need to be monitored in the long-term—like an active unit being replaced by a spare unit—or keep an eye on the status of supplies—like using the last lamp or being low on replacement gel. Since most Lighting Supervisors are responsible for a full season of productions, it is important for you to keep an eye on the status of shows that are in performance while still empowering the running technician to manage the day-to-day.

## STRIKE

The final part of the process is strike. After the show's run finishes, the Lighting Supervisor will come back in to make sure that everything gets taken down properly. Like we talked about in load-in, this is where you restore the venue to "zero." All temporary power and data infrastructure is removed, and equipment is returned to storage or to the company you rented it from. Take care to ensure that you are restoring the space to that "zero" level, especially if you are not immediately rolling over to another load-in.

At some point after the show opens, some producers will hold a "post-mortem" discussion about the project. Even if they do not, hold one for yourself. Invite the Lighting Team if they are available. Reflect on your challenges and how you handled them. It is easy to look at a completed process and say, "I made it." You did. That is great. But, tomorrow, you will start a new process and you owe it to yourself to be better. You are your own best teacher. Acknowledge that some things could have been better and figure out how to get there. As I often say to my teams—always do your best, but never stop trying to be *the* best.

# Part Three
# *Innovation*

With each passing year I look at the light plots that cross my desk and say, "I want to go back to the days of just hanging lights and plugging them in!" Light plots today are so much more than just a bunch of conventional fixtures. Almost every show, it seems, needs a custom control network or a custom light fixture. On some shows, I spend more time planning and building set electrics than I do hanging lights. Of course, as much as pretend to long for the "good old days"—which probably never existed in the first place—I relish the opportunity to solve the crazy challenges that each new show presents. If you are a Lighting Supervisor, you must be an innovator. You get so used to thinking outside the box that you forget what the box even looks like!

# CHAPTER 12

# Boy, Wouldn't It Be Cool if…?

## PUSHING THE BOUNDARIES

As a technician, the Lighting Supervisor deals with cold hard reality every day. A cable has a maximum ampacity and there is nothing you can do to change that. However, for the artistic team, dreams may seem just as real. In many ways, this is the most important part of the collaboration between a Lighting Supervisor and a Lighting Designer. The Lighting Designer will always try to push the boundaries of reality and conventionality to create a unique and memorable artistic experience. The Lighting Supervisor must figure out how to make that experience fit into the box of reality.

When you first see the scenic drawings or the light plot, there will almost always be something that makes you shout, "impossible!" While some things are impossible, the fact of the matter is that most things are not. Declaring that something is impossible is the quickest way to shut down a dialogue with the Lighting Designer and the rest of the artistic team. The process of taking a show from idea to reality is long. Maintaining good relationships is key to making it through.

When dealing with "impossible" ideas, it is important for you to remain open. What can you do to make the "impossible" possible? For example, I was once asked to make a light bulb magically light up in an actor's hand. When a designer or director uses the word "magical" it is very easy for your brain to shut down and wave the "impossible" flag. Of course, even Houdini knows that magic is just an illusion. The light bulb only needs to look like magic. So, how do I make a light bulb look like it is magically lighting up in an actor's hand?

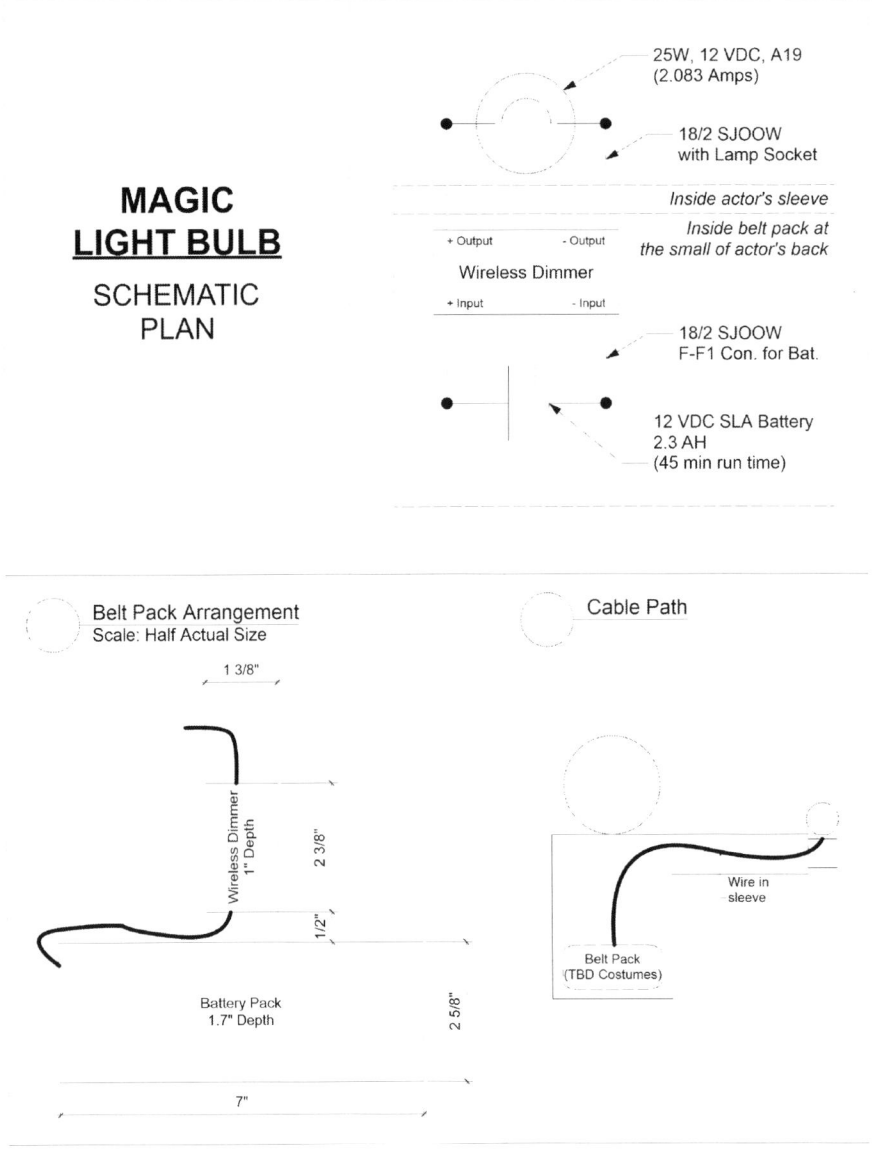

**Figure 12.1** Plan for "Magic" Light Bulb attached to the actor..

My first idea—illustrated in Figure 12.1—was to use a 12-volt light bulb. These bulbs are commonly found in recreational vehicles. With this plan, I would have the actor conceal a low-profile lamp socket in his palm and run a low gauge SJOOW cable through his sleeve to a battery and wireless dimmer in the small of his back. The costume department would be enlisted to build a waist band that would work like a mic pack.

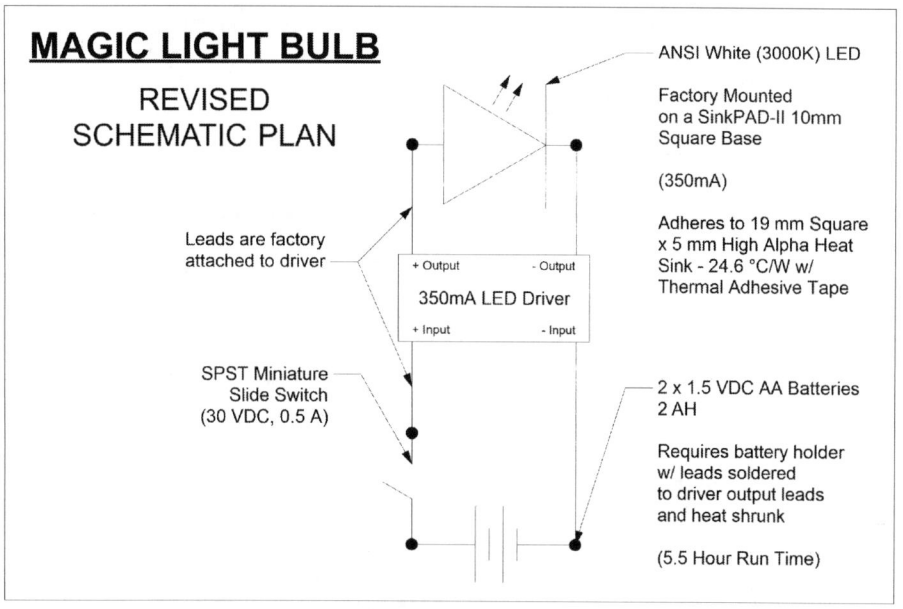

**Figure 12.2** Plan for self-contained "Magic" Light Bulb.

While this idea certainly had merit because it allowed the designer to control the bulb in the actor's hand without any visible wires, it did not meet the director's needs because they wanted the bulb to be able to be passed among multiple actors. I had not asked if multiple people were going to hold the bulb, so I did not know that when I came up with my first plan.

Back at the drawing board, I knew I was going to need to devise a project that was self-contained. The battery, control, and light source were all going to need to be on-board. This meant that the "light bulb" was going to need to be a container, not an actual light bulb. Luckily, I was able to find a novelty plastic light bulb and pack it with the necessary gear. Instead of using a manufactured light source—the R.V. bulb—I was going to need to assemble a light source from component parts. I was going to need to research light source options—as you can see in Figure 12.2, I chose a high-intensity LED—and determine what other components would be needed to power and control it. Analyzing the power requirements of the LED and the duration of time needed for their operation led me toward AA Batteries as the power source.

The two AA batteries needed to power the LED for the defined length of time were nearly the same size as the light bulb, shown in Figure 12.3. Given that size, I would need a very small wireless controller if I were to retain remote controllability. Unfortunately, such devices can be very expensive and while I was researching the project, the artistic team increased their need to ten magic light bulbs. There was not budget available to

**Figure 12.3** Mass producing "Magic" Light Bulbs.

support such a purchase. With remote control off the table, I designed the bulb to use an actor operated switch. Losing remote control was sad, but did not ruin the effect. Thus, the impossible was made possible—ten "magic" light bulbs.

This example project illustrates several steps that are required when going down a journey with an outside-the-box idea.

- Understanding the Idea
- Planning a Solution
- Budgeting the Solution
- Prototyping and Solution Approval
- Execution

## UNDERSTANDING THE IDEA

One of the keys to managing special projects—whether they are set electrics, practicals, or other ideas—is to understand what the design team is going for. I recommend asking a host of questions to try to determine what the idea is and what it will need to do.

For example, I have done many shows where the artistic team will ask for a real refrigerator on stage. My first question is always, "does it need to keep things cold?" I have found that most of the time, when the team says "real refrigerator" they mean one that lights up when you open it, not one that keeps things cold. If dealing with just a light source, it is usually better to put a dimmer controlled bulb inside rather than actually plug in the refrigerator. Refrigerator compressors can be quite loud in a theatrical setting, so if the refrigerator does not need to keep things cold, it is often better that the noise be avoided. I learned that the hard way when a furious director in tech could not stand the compressor noise. Since I did not take the time to understand what was really wanted and just plugged in a refrigerator like I was asked, I created a problem that could have been avoided.

Normally, I divide my questions into Magic and Limits. When I ask about Magic, I want to know what the artistic team wants the project to look like in the end. For me, this separates the execution of the project from its' intended effect. It is easy to get caught up in a discussion of how the effect will be made before you understand the goal of the project. Using the magic light bulb as an example, the magic is "I want the actor to hold an illuminated light bulb and pass it around the stage." You do not want to get caught up in the realism of that idea because it is not real. The solution you use will just be an illusion. I do not have to use an actual light bulb. I just need to make it look like I am.

For limits, I want to know what the project should not do. With the light bulb, I know that it cannot have any visible wires and that it cannot be physically connected to the actor. This means that whatever source and control I use will need to be fully contained in the object.

As another practical example, take the Birthday Cake that I mentioned in Chapter 1. The artistic team wanted a realistic birthday cake that would be bright enough to illuminate an actor's face. When I asked about the magic for this cake, I learned that the goals were:

- Realistic looking cake.
- Bright enough to illuminate the actor's face when the scene light is low or at zero.
- Able to be completely doused on command.

Looking into the limits I learned that:

- A single cake needs to last four weeks but will not be eaten.
- No live flame can be used.

Given those parameters, I knew we needed to build a fake cake with a remote controllable synthetic light course. In this case, the Props Department took charge of designing and building the cake itself. I provided them with technical requirements for the design—seen in Figure 12.4—such as the size of the electronics. With Props concentrating on the look, I focused on planning the light source, control, and power requirements. Props used a plastic bucket as the starting place so that the cake was completely hollow and durable. For candles, I used "grain of wheat" incandescent light bulbs that props put inside drinking straw "candles". These light sources required a 12-volt battery and were able to be controlled with a wireless dimmer.

## PLANNING AND BUDGETING THE SOLUTION

The planning of the solutions to these outside-the-box ideas falls squarely on the Lighting Supervisor and their technical colleagues. You will need to think through the solution in detail. To help with this process, I make schematics like the ones shown in Figures 12.1, 12.2, and 12.4. A simple schematic will help you virtually assemble the project and consider what you will need to complete it.

For example, all lighting projects are comprised of the same four types of components—light source, power, control, and wiring. When considering the birthday cake in Figure 12.4, I can easily see that I need the small light bulbs, a battery, and a dimmer,

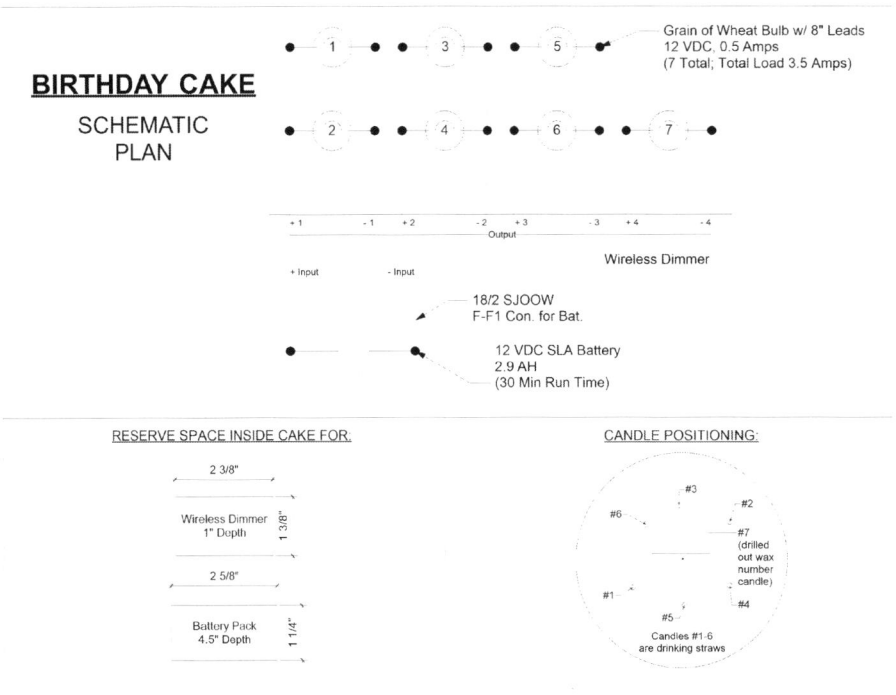

**Figure 12.4** Plan for remote-controlled birthday cake.

but if I do not draw it out, I can forget that I need wire to connect it all. This wire will have specific requirements and might not be something that I have lying around my shop.

Your schematics will want to be detailed and easy to understand, but do not need to be incredibly formal. In addition to helping you think through the project, these drawings will be how you convey your idea to other technical managers and then ultimately to the technicians who will assemble the device. I find it easiest to draw my schematics in Vectorworks, but it can be just as effective to sketch something on paper. In addition to illustrating the circuit path, I also try to provide drawings that show how things will physically fit together. This is especially important if another department will be constructing the container for the lighting device, as was the case with the birthday cake example.

As you draw your schematic, the supplies you need for your project will start to jump out. Just as you did with the larger light plot, it is important to think through each element of the project, what its' requirements are, and where those supplies are coming from. In Figure 12.5, I list each element needed for the magic light bulb project and what they cost to ensure the project will fit within budget. Note that each element is detailed. I do not just need "a battery," I need a "1.5 V.D.C. AA Battery." As I plan the project, it is important to consider what will be needed in a detailed level.

Sometimes you will be building devices from individual components. Your light source might be a light bulb, but that light bulb is comprised of the glass envelope, a filament, and a brass screw base. While in almost all circumstances, you will buy those components pre-assembled as a light bulb, but the same might not be true with other light sources like LEDs. If you purchase a component LED, you only have the diode. The LED will require heat and current management components to function properly. The supply list

| Project | Item | Source | Qty | Item Cost | Total Cost |
|---|---|---|---|---|---|
| Magic Bulb | AA Battery Holder with Leads | Amazon | 13 | $ 3.41 | $ 44.33 |
| Magic Bulb | Heat Shrink (22-18 AWG), 96" | Amazon | 1 | $ 6.43 | $ 6.43 |
| Magic Bulb | Plastic Light Bulbs, Clear (Sold as 24 Pack) | Amazon | 24 | $ 0.82 | $ 19.75 |
| Magic Bulb | Solder SAC305 WS482 3% .050 DIA | IN STOCK | 1 | $ - | $ - |
| Magic Bulb | White Spray Paint, Flat | IN STOCK | 1 | $ - | $ - |
| Magic Bulb | ANSI White (3000K) LUXEON Rebel Plus LED on a SinkPAD-II 10mm Square Base - 166 lm @ 700mA | Luxeonstar | 13 | $ 4.99 | $ 64.87 |
| Magic Bulb | 19 mm Square x 5 mm High Alpha Heat Sink - 24.6 °C/W | Luxeonstar | 13 | $ 4.35 | $ 56.55 |
| Magic Bulb | Pre-Cut, Thermal Adhesive Tape for 10 mm Square LED Assemblies - (10 Piece Sheet) | Luxeonstar | 2 | $ 3.14 | $ 6.28 |
| Magic Bulb | 350 mA MicroPuck DC Driver - With Leads | Luxeonstar | 13 | $ 9.49 | $ 123.37 |
| Magic Bulb | Assembly Press for Mounting Single Rebel LED Assemblies to a Heatsink | Luxeonstar | 1 | $ - | $ - |
| Magic Bulb | SPST Miniature Slide Switch (30 VDC, 0.5 A) | Radioshack | 13 | $ 2.50 | $ 32.44 |
| Magic Bulb | AA Batteries (1.5 VDC, 2 AH) (Sold as 20 Pack) | Target | 40 | $ 0.80 | $ 31.98 |

**Figure 12.5** Budgeting the "Magic" Light Bulb.

in Figure 12.5 shows that the magic light bulb's LED will require a heat sink, thermal adhesive, and a driver.

With the solution planned and budgeted, you will need to ensure that it is possible to be completed before pitching it to the artistic team. You may discover that your plan requires more components than you can afford within the show's budget. You may also learn that the process of executing the plan will take too many technicians or too long a period of time.

For example, Figure 1.1 in Chapter 1 shows a light plot with a large scenic element comprised of 263 light bulbs. To complete that project, the technicians will have to wire all those sockets and install all those light bulbs during the build and installation windows. If there is not the time in the schedule or enough available technicians to do that work, I need to address that before presenting my solution to the artistic team.

If an outside-the-box idea is going to be declared impossible, it is normally going to happen during the budgeting process. You can almost always develop a plan to create an illusion, but sometimes those plans become too expensive or too time consuming. You might be able to take an over-budget idea back to the drawing board and simplify it, but you might not. If you cannot fit an idea in the budget, you will have to take that back to the artistic team. Maybe the magic or the limits can be altered, and you can return to planning. This back and forth can take some time, so you need to start working on these projects as soon as you learn of them.

## PROTOTYPING AND SOLUTION APPROVAL

Your first idea may not always work, so it is good to test it out before committing to it. It is also a good idea to show the artistic team something in process so that you can ensure you are all on the same page. The further you go down a path, the more committed you are to it. Try to get the details ironed out early.

For example, Figure 12.6 is a prototype of an LED Halo effect. The design called for over 200 feet of molding with this halo emerging from under it. Thus, it was important that the artistic team was able to approve the look of the halo before the set was constructed. There was not going to be time to make a substantial change if it did not work as expected.

Additionally, taking time to do a small subset of the work yourself will help you direct the installation. You will be able to let the technicians know what things to watch out for and what techniques work better than others. Sometimes, you will discover that the idea you had does not work at all in practice and you will have to come up with a different plan or solve a challenge that you did not expect. Prototyping should happen as early in the process as possible. You should not pitch a solution to the artistic team that has not been prototyped.

**Figure 12.6** Prototyping an LED Halo for a scenic piece.

With all your planning complete, schematics and documentation prepared, and prototype in hand, you can finally pitch your solution to the artistic team. This allows them to see their outside-the-box idea in real life. If it matches their imagination, you will be able to move forward. If not, you will be well prepared to engage in a conversation about how it can change or how it cannot. Once you and the artistic team agree on a doable solution you can move forward with execution.

## THAT'S NOT THE INTENDED PURPOSE

One of the important things to consider when trying to innovate and solve problems like these is the balance between an outside-the-box idea and stretching equipment beyond its' capacity. Most theatre technicians have walked the aisles of the local hardware store looking for a solution to a problem not based on what an object was intended to do, but rather what it can do. Certainly, this is a big part of innovation, but you must be careful that you do not enter unsafe territory.

As an example, in the last chapter I mentioned using a slip ring to avoid cable twisting in a revolve. When searching for a solution to that problem, it was important that whatever equipment I used could withstand the situation I was going to put it in, even if that was not the situation it was manufactured for. I had to research all the electrical and structural capacities of the slip ring I wanted to use and devise a safe method for installing it. In this way, despite not using the device in a wind turbine—its' intended purpose—I was able to install it in a manner that fit within the requirements of the manufacturer.

Thus, the more innovative you want to be, the more stringent with practice you must become. When using a stage light, you know that the fixture has been tested and approved for your use. Its' manual will tell you what its' power and other safe use requirements are. When you are developing an outside-the-box solution, you are an inventor. You do not have a manufacturer to tell you exactly what to do. To help with this, I always assemble my projects with engineered connections and quality components. I may have to determine for myself how to safely use the final project, but if I use components with documentation, I can evaluate the safe use of each component individually and assemble something that I know will be reliable.

For example, battery capacity is a complicated concept. Most batteries are rated in amp-hours. Without any additional research, you can assume this rating is simply amps per hour. However, this rating refers to a specific condition. It is the duration, in hours, the battery will supply exactly 1 Amp of current before completely being drained. While this number does tell you a battery's capacity, it cannot be used at face value. To calculate how long your batteries will last, you will need to use Peukert's formula and know the battery's Peukert number—a measure of its' efficiency. Peukert's formula looks like this:

$$T = C \div I^n$$

"$T$" is the amount of time until 100% drainage, "$C$" is the listed capacity of the battery, "$I$" is the current of the load in Amps, and "$n$" is the battery's Peukert number, normally found in the manufacturer's battery specifications. A little math tells you that a 10 Amp-hour battery with a two-Amp load and a 1.2 Peukert number would be complete discharged in 4.35 hours.

However, most lead acid manufacturers recommend that a battery only discharge 80% of its' capacity so that it can be recharged. This means that you can only use 80% of the available 4.35 hours or 3.48. If a production had a running time of 2 hours, you would need to make sure a freshly charged battery was used for each production to ensure reliability.

Innovating and inventing does not preclude you from understanding how the components work. In fact, because you are creating a new use condition, you are more responsible than ever for understanding each components' requirements.

My favorite example of failing to consider this comes from a retrofit project that I worked on. The goal was to retrofit a sodium vapor fixture with an incandescent source of similar brightness. My retrofit was well-executed and electrically sound, but I did not carefully consider the heat rating of the fixture's enclosure. In a short amount of time, the housing had melted like an ice cream cone on a hot day.

Experimentation and innovation must be done with caution and with research. Start each process early so that you can develop a project that delivers the result you want. Ask questions, research components, and test theories. If you determine that a designer's dream project is not doable, that is okay too. It is tempting to try to do something without the right equipment or without the right expertise just to satisfy the artistic team, but too often those projects result in unsafe or unreliable situations that are much worse than simply not having the desired effect.

CHAPTER 13

# Asset Management and Season Planning

## LOVING YOUR GEAR

Lighting and Sound departments are unique among their production peers because instead of building or purchasing new items for each production—like is common with Scenery, Costumes, and Props—they need to maintain the same equipment for multiple years' worth of shows. This can make understanding the annual costs of a Lighting or Sound department quite different than that of their peers. Most of a scenic budget, for example, will go towards wood, steel, and hardware. These supplies are visible in each production. If a production is small, expenses go down. If a production is large, expenses go up. For the Lighting Department, the biggest expenses will be in equipment repair and maintenance. The amount of equipment the producer owns and needs to maintain remains relatively constant from year to year. These costs can be hidden—especially for your producer's accountant—so it is up to the Lighting Supervisor to know their gear and know what it takes to keep it performing year after year.

Professional-quality theatrical lighting equipment can be expensive particularly in the quantities that most producers require it. A simple 100-unit light plot can easily contain $50,000 worth of equipment. Even renting that much equipment would put costs in the thousands. Most producers cannot afford to incur those sorts of expenses on every production or even annually, so equipment must be made to last years or even decades.

As Lighting Supervisor, you need to know each fixture in your inventory intimately and must work to keep those fixtures in top notch condition. In Chapter 6, I recommended that you save 10% of your inventory as spare to cover any emergencies. This means that at least 90% of your inventory needs to be fully functional at any given moment. Therefore, it is particularly important to develop an effective maintenance plan for your equipment.

When developing your plan, balance the manufacturer's recommended maintenance schedule with the available time in yours. While the goal is always to perform maintenance at the frequency the manufacturer recommends, it is not always practical. It is important to determine a realistic schedule that honors the need for maintenance without stretching you and your team beyond their capacity. For example, most regional theatres take a summer hiatus. This is a great time to perform maintenance on equipment, however, it is often not financially viable for the producer to employ the full staff while the stages are dark. Consequently, there may not be adequate hands to perform maintenance during the summer. Instead, you may need to perform maintenance on some equipment at the top of the new season and on others at the end. You may not even be able to do all the equipment every year.

While it is certainly best to perform maintenance on all your equipment at least annually, there is little value in frustrating yourself about not completing it every year. Instead, determine a method for which maintenance can be tracked and completed over multiple years. If you only have capacity to maintain one third of your equipment in a year, determine what equipment, if any, must be maintained annually and then put other equipment on a tri-annual cycle—year one might be spot fixtures, year two might be wash fixtures, and so on. It is better to perform quality maintenance over a longer period than to rush through a process just to meet a desired frequency.

In addition to performing routine maintenance, you want to ensure that you have the capacity to perform repairs should they be required. The goal of maintenance, in general, is to correct any issues before they become repairs. However, even when maintaining equipment regularly, equipment will eventually fail. When this happens, it is important to have a plan in place. In some instances, you will be able to repair the equipment yourself. In other situations, you will need to send equipment to a repair service.

As Lighting Supervisor, you should be constantly evaluating your repair capacity. First, consider your staff, including yourself. What skills do you have as part of your in-house repair team? What skills do you lack? There is nothing wrong with having gaps in your team's skill set if you can develop a method for bridging those gaps.

For example, you may have a team of talented technicians, but none that can perform PCB-level repair. With this understanding it will be important for you to establish a budget for sending equipment to off-site repair companies when PCB-level repairs are necessary. When evaluating these costs consider how they compare to recruiting staff with these missing skills. If you live in a remote part of the country, sending equipment for repair can get expensive. It may be more affordable in the long term to have someone on site with those skills.

However, if you plan to rely on staff to do repairs consider how the time they spend doing that work may detract from their availability to perform their other duties. You may hire a brilliant repair technician, but also need them to work on load-ins. Do they have enough time to do repairs and load-ins? Are you paying for an in-house repair technician, but not giving them time to work on repairs?

Second, consider your tools and parts. Repairing equipment with the appropriate tools and parts is much easier than attempting to do so without them. I spent way too long trying to replace driver chips with small screwdrivers and pliers before learning the effectiveness of a chip puller. In many ways, knowing what parts or tools you need will come from the same experience that brings you the skills to complete the work in the first place. However, manuals and online trainings can provide tremendous help.

## ASSET TRACKING

The more equipment your department owns, the more complex it will be to manage it. A small black box venue might have 100 ellipsoidal spotlights. They probably never need to leave the room and tracking them can be done simply by counting and comparing to a list you keep on your clipboard. However, as you grow your career to bigger venues and multi-venue organizations, it will become paramount that you develop a system of asset tracking that scales with you. A mentor of mine once encouraged me to develop scalable workflows. This way when I found myself in a bigger place, I did not need to reinvent the wheel. What works in big venues works in small venues, but some things that work in small venues can struggle to meet the needs of big venues. Asset tracking is a good example of this.

A good asset tracking system will be able to readily help you do each of the following.

- Know the quantities of equipment that the company has available for various projects.
- Track what equipment is currently used and what equipment is available.
- Record information about the equipment's repair and maintenance history.
- Preserve details about the value and age of the equipment to track its depreciation over time.

There are many ways to go about tracking this information and the level of granularity you can have will depend on the type of method you use. The simplest method would be to keep lists, either digitally or on paper, and supplement them with any additional documentation like purchase or repair invoices and fixture counts from current shows. In this method, you can have a master list of all of the equipment owned by the producer, a list of available equipment for use by a designer with your 10% spare taken out, and another list regarding history and depreciation information.

While this method can be effective for smaller quantities of equipment, it can get out of hand when you try to scale it. For a more scalable option, I recommend that you use a relational database, like the one shown in Figure 13.1, to track your assets. Relational databases come in many forms—Microsoft Access, Claris File Maker, Airtable, to name a few. Each software solution comes with different advantages and challenges. Select a system that you feel that you can master given the amount of time you have to devote to it. I like to use Microsoft Access. This is mostly because it was already installed on the many of the computers I work with regularly, thus not incurring any additional expenses.

Figure 13.1 Microsoft Access inventory database

Relational databases are excellent tools because they take the lists you were making and links them. For example, if you track nonfunctioning equipment—often abbreviated "NFG"— in a digital repair queue, you can use that data to produce an inventory sheet for the designer that hides any NFG's.

Additionally, relational databases allow the generation of forms, reports, and a variety of other paperwork that make the entry and viewing of information clear and easy. One list of information can be queried and interacted with in a myriad of ways. For example, if you record NFG data to track a repair queue, you can also decide to keep that data as a log. Should the same piece of equipment break again, you can generate a report of its' repair history and potentially speed up its' next diagnosis.

Regardless of which method you use, look to ensure you are recording comprehensive details and avoid using jargon or nicknames. For example, if you have an entry for an "ETC Leko," that can be very misleading. It may be common for some technicians or designers to refer to all ellipsoidal spotlights as "Lekos," but this does not help someone understand what specific fixture you own. ETC's flagship ellipsoidal is called the "Source Four." Further, ellipsoidal spotlights come in a variety of sizes with a variety of features. None of that is clear when you just say "ETC Leko." A better entry may be, "ETC Source Four, Black, with 36-degree Lens Tube." Now any reader can know exactly what the entry refers do.

If you have a staff, also remember that they may need to read and edit this documentation too. Consider how new equipment is added to your inventory and how others will interface with it. If using a relational database, for example, you may make a "New Equipment" form. Any one of your department staff members can fill out a digital form that automatically updates the inventory sheet that you give to your designers.

Finally, it is important to track the depreciation and value of your equipment. Depreciation is a complex concept. The goal is to predict how much your equipment will be worth in the future. A new piece of equipment has a certain value. That value is connected to both its' newness and its innovativeness. As of this writing, Philips Vari*Lite's VL2600 is a new piece of equipment. It is innovative in its ability to use LED technology to do the work previously thought only possible with incandescent or arc-source moving lights. The current value of a fixture right off the line is the highest that fixture is likely to ever be worth. Even if the fixture is kept in its' box and never used, its' value will still decrease over time because technology will improve around it. If the fixture is used, its' value will decrease faster due to the wear and tear on its' parts.

To calculate depreciation, start with the amount a new unit currently costs and the amount of time a new unit will last before the cost of repairing will be too high. The later number, referred to as its' "useful life," is always a bit of a guess dependent on your tolerance for risk. If you have a low tolerance for risk, you may use the amount of time until the piece of gear will need its' first repair. If you have a high tolerance for risk, you may estimate how long it will be before the equipment is discontinued. When first evaluating useful life, do some research on similar products. How often does a new version of the product come out? How long has the product existed and how ubiquitous is it?

As an example, the ETC Source Four debuted in 1992 and, as of this writing, it is still the most common lighting fixture of its' kind in the industry. Most repairs on the unit can be done by technicians with moderate training. Common parts are affordable. As the industry now starts to move beyond incandescent sources, the end for this fixture may be in sight. Therefore, I might say that it has a useful life of 20 to 30 years. Meaning that a unit would be "worthless" 20 to 30 years after its' initial purchase. Consequently, if I purchased one in 2005 for $500, its' current value is likely $250 because it is halfway on its' path to depreciation. Additionally, I would want to plan to replace the fixture before it was fully depreciated in 2035.

## CAPITAL EXPENSES

Using an understanding of depreciation, you can evaluate your department's capital needs. In the simplest terms, capital expenses can be considered expenses used to maintain the current level of equipment and technology and mitigate the effects of depreciation. To understand where your department's inventory may lie, estimate depreciation for your units and determine what year each will become fully depreciated like the Source Four example we just looked at. Use this information to create a depreciation schedule so that you can see by which year different items will need to be replaced.

With your schedule complete, you can determine what type of fixture you would expect the approximate replacement for a piece of equipment would be. For example, if I had Source Fours that were purchased in 1992 and I determined that they would fully depreciate in 2022, I would recommend that they be replaced by a newer Source Four. In 2022, it seems likely that the common replacement would be a newer version of itself. However, if I expected to replace the units in 2035, I might imagine that LED fixtures would likely dominate the market and would suggest that the capital expense be LED fixtures.

With an understanding about potential replacements, you can begin to assign dollar values per year as illustrated in Figure 13.2. In doing so, you can see if particular years will require more substantial investment than other years and plan to spread the investment over multiple years. Using the example above, if you had 100 fixtures that were needed to be replaced by LEDs in 2035, it could be expected that the cost would be near $250,000. This may be challenging to manage. Instead, you could plan to purchase six or seven units per year over 15 years and spread that cost out while still ensuring that the total fixture count would be replaced by the designated year.

It is important to continuously evaluate, research, and revise your capital projections. Try to dedicate money to reinvestment as frequently as possible. Small investments are generally much easier to manage than large ones, so do your best to always think

| Items | FY2020 | FY2021 | FY2022 | FY2023 | FY2024 | FY2025 | FY2026 | FY2027 | FY2028 |
|---|---|---|---|---|---|---|---|---|---|
| **Mainstage Ellipsoidals** 100 Units Depreciation in 2027 | | | | | $10,000 20 Units | $10,000 20 Units | $17,500 35 Units | $12,500 25 Units | |
| **Mainstage Washes** 50 Units Depreciation in 2025 | | | | $2,500 10 Units | $5,000 20 Units | $5,000 20 Units | | | |
| **Studio Ellipsoidals** 25 Units Depreciation in 2027 | | | | | | | $7,500 15 Units | $5,000 10 Units | |
| **Moving Lights** 10 Units Depreciation in 2023 | $15,000 3 Units | $10,000 2 Units | $15,000 3 Units | $10,000 3 Units | | | | | |
| **Mainstage Console** Depreciation in 2021 | | $10,000 New Console | | | | | | | |
| **Studio Console** Depreciation in 2028 | | | | | | | | | $10,000 New Console |
| **FY TOTALS** | $15,000 | $20,000 | $15,000 | $12,500 | $15,000 | $15,000 | $25,000 | $17,500 | $10,000 |

**Figure 13.2** Capital expense chart.

about what your organization will need in ten or fifteen years. If an opportunity arises to purchase equipment for a show instead of renting it, use your understanding of your company's capital position to make that determination.

## SEASON BUDGETING

In addition to capital expenses, a Lighting Supervisor will need to plan for their annual department expenses. These will be a combination of general costs and production costs. When evaluating general costs, consider your capital, maintenance, and repair needs. I recommend using the maximum amount of specificity that you can manage when considering these costs. For example, in Figure 13.3, I do not just list $5,000 for supplies. I break it down—$250 for connectors, $25 for electrical tape, $250 for gaff tape, and so on.

Having more specific detail has two clear benefits. First, you will be able to understand and respond to the specific impact of any budget cuts. If I request $5,000 and am told I can only have $4,800, I may not immediately understand the impact of that. Whereas, if I were more specific, I could understand that may mean that I would need to be less liberal with my gaff tape usage or not purchase a new cable tester.

Second, having that specificity improves your ability to discuss your budget in the first place. While it is common for a Lighting Supervisor to make a budget request, they will almost never be the person deciding if they can have that money. The more specificity

| Account No. | Account Description | ExpCd | Line Item | FY2019 Estimate | Notes |
|---|---|---|---|---|---|
| 517-01-001 | Office Supplies | OIT | Ink and Toner | $ 30.00 | |
| 517-01-001 | Office Supplies | OPP | Paper | $ - | |
| 517-01-001 | Office Supplies | OXX | Pens, Pencils, and Other Supplies | $ 300.00 | |
| 517-01-001 | Office Supplies | OSH | Sharpies | $ 25.00 | |
| 517-01-001 | Office Supplies | ONT | Name Tags and Signage | $ 25.00 | |
| 517-01-002 | Postage | TUS | Postage | $ 10.00 | |
| 517-01-002 | Postage | TXX | Fed-Ex, UPS, Etc. | $ 15.00 | |
| 517-01-003 | Phone Long Distance | HXX | General Long Distance | $ - | |
| 517-01-004 | Printing | OPT | Copies | $ 250.00 | Add this account to break out all the copy charges |
| 517-01-006 | Training/Education | NCF | Conference and Training Events | $ 675.00 | Entertainment Electrics Training or VL Training for ESM |
| 517-01-006 | Training/Education | NXX | In-house Training and Skills Labs | $ - | |
| 517-01-007 | Per Diem | DXX | General Per Diem | $ 300.00 | Training per diem |
| 517-01-008 | Travel | VDI | Distance Travel | $ 400.00 | Training Travel |
| 517-01-008 | Travel | VRN | Rental Pick-up | $ - | |
| 517-01-008 | Travel | VXX | Local Travel | $ 75.00 | |
| 517-01-009 | Lodging | GXX | General Lodging | $ 500.00 | Training lodging |
| 517-01-010 | Miscellaneous | IML | Department Morale | $ 150.00 | |
| 517-01-010 | Miscellaneous | IXX | Real Miscellaneous | $ 50.00 | |
| 517-01-011 | Dues and Fees | FXX | General Dues and Fees | $ - | |
| 517-01-050 | Equipment and Supply | ESW | Software | $ - | |
| 517-01-050 | Equipment and Supply | ERC | Improvement Projects | $ 700.00 | |
| 517-01-050 | Equipment and Supply | ESG | New Equipment for Stage | $ 400.00 | Small Objects (no Net3 Gateway - $1,600) |
| 517-01-050 | Equipment and Supply | ESP | New Equipment for Shop/Booths | $ 200.00 | |
| 517-01-050 | Equipment and Supply | ESQ | New Equipment from Wish List | $ 505.00 | SOCO Tester, XLR Tester, CAT-5 Tester (No Swisson Tester) |
| 517-01-050 | Equipment and Supply | EST | Storage Solutions | $ 400.00 | Lamp Shelf |
| 517-01-050 | Equipment and Supply | EXX | Equipment General | $ - | |
| 517-01-050 | Equipment and Supply | SCN | Connectors | $ 250.00 | |
| 517-01-050 | Equipment and Supply | SET | Electrical Tape | $ 25.00 | |
| 517-01-050 | Equipment and Supply | SGT | Gaff Tape | $ 250.00 | |
| 517-01-050 | Equipment and Supply | SNT | Splices, Wire Shoes, and Wire Nuts | $ 200.00 | |
| 517-01-050 | Equipment and Supply | SST | Spike Tape | $ - | |
| 517-01-050 | Equipment and Supply | STZ | Tie Line and Zip Ties | $ 100.00 | |
| 517-01-050 | Equipment and Supply | SWI | Wire | $ 50.00 | |

**Figure 13.3** Season budget detail.

you have in your request, the clearer the narrative around it will be. If you request $250 for gaff tape, like in Figure 13.3, be able to describe how you came to that number. Talk about how much gaff tape you used each of the last five years and how you expect next year to compare.

One of the important things regarding budgeting is that your requests can only be as specific as your records. That is to say, if you record everything you buy under the same category heading, it will be impossible to predict any cost moving forward. In order to know how much gaff tape I used over the last five years, I will have to have those records. Be sure to record your purchases with the same level of specificity you wish to use in budgeting.

In addition to general expenses, you will also need to predict the expenses that you will have for future productions. If you have no information about what a future production is, you will have to use generic expense projections. Break each production up into expense categories. I use color, templates, rentals, materials, and special effects. Find the median value per production for each expense type over the last five to ten years.

Using the median here is helpful because you can eliminate shows that had uncommon expenses from skewing your projection. With all your median figures calculated, you can add them up to determine the total amount you expect to spend on a generic production.

Using the generic numbers as a guide, you can adjust your projections based on any information you learn. If you learn that one of the generic productions will be a large musical, you will likely want to reserve more money. If the data easily exists, you can recalculate your median figures including only productions that were large musicals and get a better projection.

Of course, tracking all that data may be challenging. As a general rule, I build a sliding scale. In the middle is the general median, and on the ends are my general minimum and general maximum. It is rare that a project will set a new maximum, so you can likely use that general maximum number as a ceiling. For example, if my general color minimum is $50, median is $200, and maximum is $1,000 and I am evaluating this large musical, I can project that the color expense will be medium-high or halfway between the median and the maximum general calculations. In this case, that yields an estimate of $600.

# CHAPTER 14

# Your Turn

## RULES OF THUMB

The role of Lighting Supervisor is multifaceted and complex. They are an integral part of the production process and their significance cannot be understated. A good Lighting Supervisor in as invaluable as bad one is harmful. The most important thing you need to remember is that being a good Lighting Supervisor is not just about being experienced or knowledgeable. Sure, those things can help, but the most important thing is being self-aware enough to know your strengths and know when you need help.

Throughout this book I have provided guidance from my experience. I have put forth a lot of rules of thumb that have held true in my experience: 10% spare mitigates risk, specificity yields flexibility, empowering and teaching staff results in productivity, everyone has three hats, and others. I hope that you will consider them in your own work, but most of all I hope that you will develop your own rules of thumb.

I sat down to write this book because I wanted to lift up this role and encourage people to invest in themselves in becoming great Lighting Supervisors. Not because they wanted a steppingstone to a design career, but because they knew that being a Lighting Supervisor had value in and of itself. As you move forward in your career, help to keep this field growing. Take early-career technicians under your wing and share your rules of thumb. Write your own book!

May you always remain a learner and a teacher. May you always push yourself to be the best.

## FURTHER READING

At many times in the book, I have indicated important areas of knowledge that I felt that Lighting Supervisors should have. Here are some great resources to continue that reading.

- *Vectorworks for Entertainment Design*, Kevin Lee Allen
- *Electricity for the Entertainment Electrician & Technician*, Richard Cadena
- *Stage Rigging Handbook*, Jay O. Glerum
- *Rigging Math Made Simple*, Delbert L. Hall
- *American Electrican's Handbook*, Terrell Croft and Frederic Hartwell
- *Show Networks & Control Systems*, John Huntington
- *The Automated Lighting Programmer's Handbook*, Brad Schiller
- *The Health and Safety Guide for Film, TV, and Theatre*, Monona Rossol

# Appendix–
# Example Production Paperwork

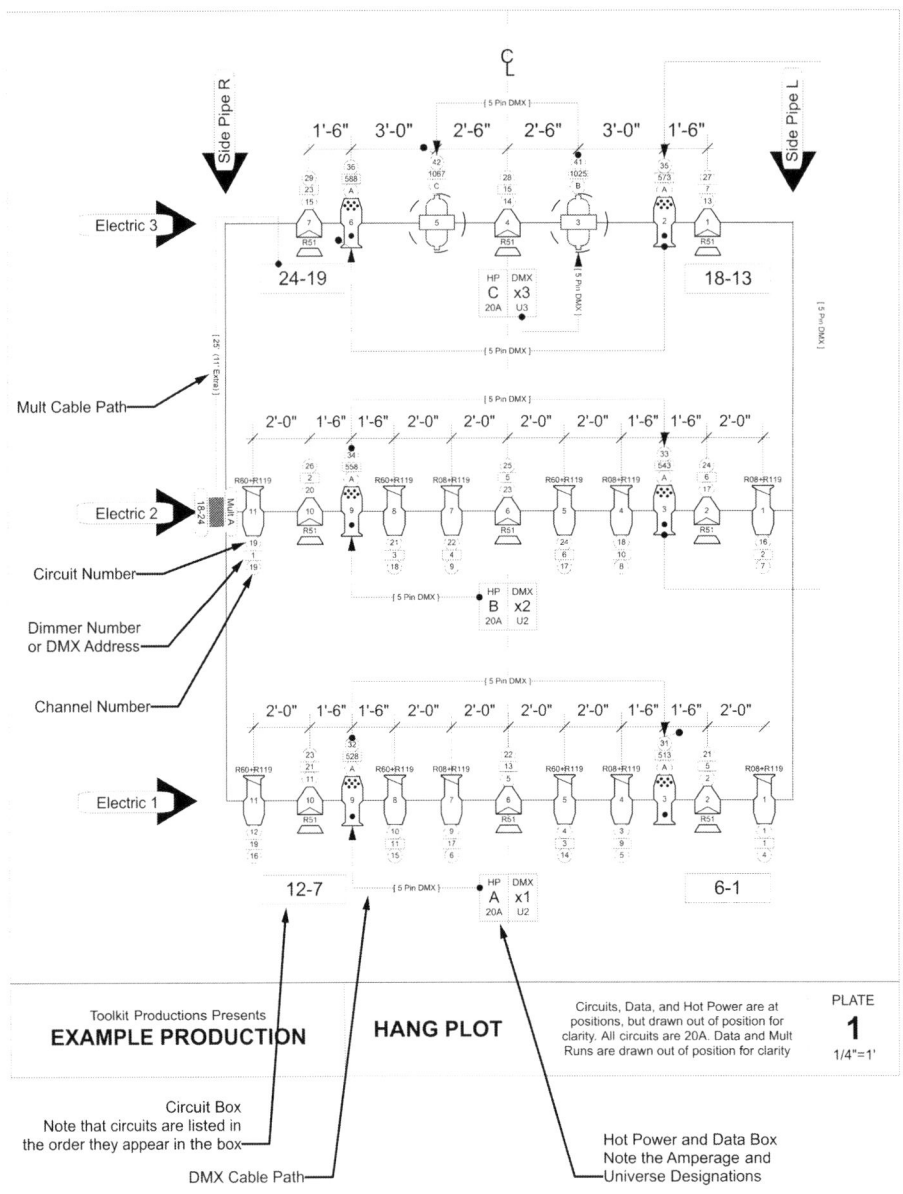

**EXAMPLE PRODUCTION**　　　　　　　　　　　　　　　　　　　　　　　　　　　　*CHANNEL HOOKUP*

Lighting by Jason E. Weber

| Channel | Addr | Ckt#/HP | Purpose | Position & Unit# | Instrument Type & Accessory & Load | Color | Gobo |
|---|---|---|---|---|---|---|---|
| (4) | 1/1 | 1 | F/L < | Electric 1 #1 | ETC S4-26 1kW | R08+R119 | |
| (5) | 1/9 | 3 | F/L < | Electric 1 #4 | ETC S4-26 1kW | R08+R119 | |
| (6) | 1/17 | 9 | F/L < | Electric 1 #7 | ETC S4-26 1kW | R08+R119 | |
| (7) | 1/2 | 16 | F/L < | Electric 2 #1 | ETC S4-26 1kW | R08+R119 | |
| (8) | 1/10 | 18 | F/L < | Electric 2 #4 | ETC S4-26 1kW | R08+R119 | |
| (9) | 1/4 | 22 | F/L < | Electric 2 #7 | ETC S4-26 1kW | R08+R119 | |
| (14) | 1/3 | 4 | F/L > | Electric 1 #5 | ETC S4-26 1kW | R60+R119 | |
| (15) | 1/11 | 10 | F/L > | Electric 1 #8 | ETC S4-26 1kW | R60+R119 | |
| (16) | 1/19 | 12 | F/L > | Electric 1 #11 | ETC S4-26 1kW | R60+R119 | |
| (17) | 1/6 | 24 | F/L > | Electric 2 #5 | ETC S4-26 1kW | R60+R119 | |
| (18) | 1/3 | 21 | F/L > | Electric 2 #8 | ETC S4-26 1kW | R60+R119 | |
| (19) | 1/1 | 19 | F/L > | Electric 2 #11 | ETC S4-26 1kW | R60+R119 | |
| (21) | 1/5 | 2 | Bax | Electric 1 #2 | ETC S4-PAR WFL+7.5" BD 575w | R51 | |
| (22) | 1/13 | 5 | Bax | Electric 1 #6 | ETC S4-PAR WFL+7.5" BD 575w | R51 | |
| (23) | 1/21 | 11 | Bax | Electric 1 #10 | ETC S4-PAR WFL+7.5" BD 575w | R51 | |
| (24) | 1/6 | 17 | Bax | Electric 2 #2 | ETC S4-PAR WFL+7.5" BD 575w | R51 | |
| (25) | 1/5 | 23 | Bax | Electric 2 #6 | ETC S4-PAR WFL+7.5" BD 575w | R51 | |
| (26) | 1/2 | 20 | Bax | Electric 2 #10 | ETC S4-PAR WFL+7.5" BD 575w | R51 | |
| (27) | 1/7 | 13 | Bax | Electric 3 #1 | ETC S4-PAR WFL+7.5" BD 575w | R51 | |
| (28) | 1/15 | 14 | Bax | Electric 3 #4 | ETC S4-PAR WFL+7.5" BD 575w | R51 | |
| (29) | 1/23 | 15 | Bax | Electric 3 #7 | ETC S4-PAR WFL+7.5" BD 575w | R51 | |
| (31) | 2/1 | A | LED Temps | Electric 1 #3 | ETC S4 Lustr2-36 171w | | G705-A |
| (32) | 2/16 | A | LED Temps | Electric 1 #9 | ETC S4 Lustr2-36 171w | | G705-A |
| (33) | 2/31 | A | LED Temps | Electric 2 #3 | ETC S4 Lustr2-36 171w | | G705-A |
| (34) | 2/46 | A | LED Temps | Electric 2 #9 | ETC S4 Lustr2-36 171w | | G705-A |
| (35) | 2/61 | A | LED Temps | Electric 3 #2 | ETC S4 Lustr2-36 171w | | G705-A |
| (36) | 2/76 | A | LED Temps | Electric 3 #6 | ETC S4 Lustr2-36 171w | | G705-A |
| (41) | 3/1 | B | ML | Electric 3 #3 | VL-2600 Profile 820w | | |
| (42) | 3/43 | C | ML | Electric 3 #5 | VL-2600 Profile 820w | | |

- Position and Unit Number
- Designer's Purpose
- Circuit Used
- DMX Address of Dimmer or Fixture
- Instrument Type, Accessories, and Wattage
- Color and Templates
- Channel Number from Designer

## EXAMPLE PRODUCTION

### INSTRUMENT SCHEDULE
Lighting by Jason E. Weber

| Unit# | B/B | Instrument Type & Load | Purpose | Color | Gobo | Access | Mul | Leg# | Cabling | C#/HP | Addr | Chan |
|---|---|---|---|---|---|---|---|---|---|---|---|---|
| **Electric 1** | | | | | | | | | | | | |
| 1 | | ETC S4-26 1kW | F/L < | R08+R119 | | | | | | 1 | 1/1 | (4) |
| 2 | | ETC S4-PAR WFL 575w | Bax | R51 | | 7.5" BD | | | | 2 | 1/5 | (21) |
| 3 | R | ETC S4 Lustr2-36 171w | LED Temps | | G705-A | | | | | A | 2/1 | (31) |
| 4 | | ETC S4-26 1kW | F/L < | R08+R119 | | | | | | 3 | 1/9 | (5) |
| 5 | | ETC S4-26 1kW | F/L > | R60+R119 | | | | | | 4 | 1/3 | (14) |
| 6 | | ETC S4-PAR WFL 575w | Bax | R51 | | 7.5" BD | | | | 5 | 1/13 | (22) |
| 7 | | ETC S4-26 1kW | F/L < | R08+R119 | | | | | | 9 | 1/17 | (6) |
| 8 | | ETC S4-26 1kW | F/L > | R60+R119 | | | | | | 10 | 1/11 | (15) |
| 9 | R | ETC S4 Lustr2-36 171w | LED Temps | | G705-A | | | | | A | 2/16 | (32) |
| 10 | | ETC S4-PAR WFL 575w | Bax | R51 | | 7.5" BD | | | | 11 | 1/21 | (23) |
| 11 | | ETC S4-26 1kW | F/L > | R60+R119 | | | | | | 12 | 1/19 | (16) |
| **Electric 2** | | | | | | | | | | | | |
| 1 | | ETC S4-26 1kW | F/L < | R08+R119 | | | | | from E3 | 16 | 1/2 | (7) |
| 2 | | ETC S4-PAR WFL 575w | Bax | R51 | | 7.5" BD | | | from E3 | 17 | 1/6 | (24) |
| 3 | R | ETC S4 Lustr2-36 171w | LED Temps | | G705-A | | | | | A | 2/31 | (33) |
| 4 | | ETC S4-26 1kW | F/L < | R08+R119 | | | | | from E3 | 18 | 1/10 | (8) |
| 5 | | ETC S4-26 1kW | F/L > | R60+R119 | | | A | 6 | via Mult | 24 | 1/6 | (17) |
| 6 | | ETC S4-PAR WFL 575w | Bax | R51 | | 7.5" BD | A | 5 | via Mult | 23 | 1/5 | (25) |
| 7 | | ETC S4-26 1kW | F/L < | R08+R119 | | | A | 4 | via Mult | 22 | 1/4 | (9) |
| 8 | | ETC S4-26 1kW | F/L > | R60+R119 | | | A | 3 | via Mult | 21 | 1/3 | (18) |
| 9 | R | ETC S4 Lustr2-36 171w | LED Temps | | G705-A | | | | | A | 2/46 | (34) |
| 10 | | ETC S4-PAR WFL 575w | Bax | R51 | | 7.5" BD | A | 2 | via Mult | 20 | 1/2 | (26) |
| 11 | | ETC S4-26 1kW | F/L > | R60+R119 | | | A | 1 | via Mult | 19 | 1/1 | (19) |
| **Electric 3** | | | | | | | | | | | | |
| 1 | | ETC S4-PAR WFL 575w | Bax | R51 | | 7.5" BD | | | | 13 | 1/7 | (27) |
| 2 | R | ETC S4 Lustr2-36 171w | LED Temps | | G705-A | | | | | A | 2/61 | (35) |
| 3 | R | VL-2600 Profile 820w | ML | | | | | | | B | 3/1 | (41) |
| 4 | | ETC S4-PAR WFL 575w | Bax | R51 | | 7.5" BD | | | | 14 | 1/15 | (28) |
| 5 | R | VL-2600 Profile 820w | ML | | | | | | | C | 3/43 | (42) |
| 6 | R | ETC S4 Lustr2-36 171w | LED Temps | | G705-A | | | | | A | 2/76 | (36) |
| 7 | | ETC S4-PAR WFL 575w | Bax | R51 | | 7.5" BD | | | | 15 | 1/23 | (29) |

Mult Name and Leg Number
Note to help technicians find a circuit
Circuit Name or Number
Note that circuit numbers are different than dimmer addresses.
This exemplifies working in a "Hard Patch" environment
Denotes Rental Gear

EXAMPLE PRODUCTION PAPERWORK 181

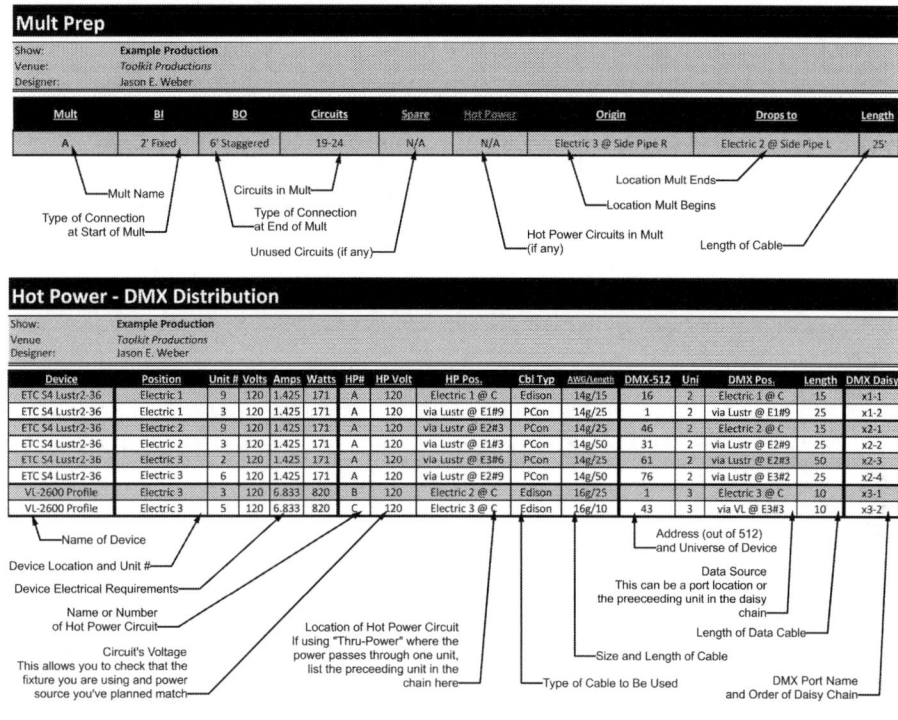

# Index

Note: *Italic* page numbers refer to *figures*.

## A

ampacity 75–77, 80
architect 3, 19, 21, 24; Lighting Designer as 19–21
Artistic Director 9
artistic team 153, 155, 157, 160, 161, 163
asset management 165–173
asset tracking 167–170
Assistant Lighting Designers 7, 13, 14
Assistant Lighting Supervisor 38
Associate Lighting Designers 13, 14, 132

## B

boundaries, pushing 153–156

## C

cable's ampacity 75–77
capital expenses 170–171
circuit boxes 69, 74, 124
circuit load 77–79
circuit numbers 70, 71, 85, 95, 100, 102, 122, 128
circuit planning 67, 68, 86, 87, 94
collaboration 3; *see also* peer to peer collaboration
Color and Template Prep 104–110, 113
Color Count Sheet 105, 106, 108
Color Frame Label 105, 108
contingency time 118
crew, empowering 146–148
cross-departmental projects 23, 48, 49, 51

## D

daisy-chain 90–92, 96, 97
data paperwork 94–97
Department Hat 32–33, *32*
designer inventory layout *46*
Designer-Supervisor relationship 12
design team 47–49, 157
dimmer numbers 71, 72, 85, 95, 102
Dimming & Control dialogue 81
Director of Photography (D.P.) 7–8
Director of Production 9, 12
DMX Footprint 90
DMX protocol 90, 92, 94
DMX Termination 94

## E

electrical diversity *79*
electrical planning 67–97; circuit and dimming infrastructure 67–80; hot power and data paperwork 94–97; plot dimmering 80–86; power and data infrastructure 86–94
electrical trades 5
ellipsoidal spotlights 167, 169
empowerment 27–30
engineer, Lighting Supervisor as 21–24
Entertainment Services and Technology Association (E.S.T.A.) 36
Entertainment Technician Certification Program (E.T.C.P.) 36
error checking *86*
ETCNet 92
executive leadership 9

## F

filming process 7
Final Design Meeting 49, 50
fixtures 21, 23, 55, 64, 75, 78, 83, 96, 105, 110, 114, 120, 124, 126, 127, 138, 139, 145, 169, 170

focus 131–139; calling 132–136; preparing for big game 131–132; technician focus standards 137–139
focus slack 83, 123, 124, 126
Followspot Operator 16–17
Ford, Henry 33

# G

grounder 34, 127
grounding 33–34

# H

hang cards 99, 101, 102, 104
hang wave 119–121
Hard Patch 70–72, 81, 85, 128
Head Electrician 6–8, 14, 16
hot power 86–89, 94–97, 100
Hot Power Circuit 94, 95, 126

# I

idea, understanding 157–158
independent together 33
innovation 162, 163
innovator 151
installation 23, 30, 50, 51, 53, 61, 62, 65, 67, 68, 111, 112, 115, 120, 121
Instrument Maintenance *59*
instrument schedule 100, 104, 110, 126, 128, 136, 148
International Alliance of Theatrical Stage Employees (I.A.T.S.E.) 7

# J

Jesus Bolt 138

# K

knowledge 36

# L

Label Legends 59, 60
Label Legends Manager *60*
Labor Price Out 63
leaders: are teachers 37–38
leadership: is empowerment 27–30
League of Resident Theatres (L.O.R.T.) 8

learning expert 35–37
learning relationships 38–39
Light Board Operator 16, 146
Light Board Programmer 16
Lighting Department 11, 12, 24, 27–28, 38, 48, 51, 61, 113, 118
lighting design 4–6, 12, 14, 20, 50, 67, 131, 132, 135
Lighting Designer 3–5, 7, 12–14, 16, 19–21, 24, 49, 50, 65, 129, 131, 143, 146, 153; as architect 19–21; connecting with 12–13
Lighting Director 6, 7
Lighting Engineer 35
lighting process 3
lighting team 3, 13–17, 27, 33, 50, 120, 149; two sides of *15*
Lighting Technicians 14, 16, 28, 38, 63, 65, 68, 71, 95, 96, 111, 131, 133
Lightning Tapes 103
Light Plot 4
Lightwright File 56, 57, 144
Load-in Crew 16
load-in documentation 99–104
load-in process: accessories wave 126; circuiting 121–126; color and templates wave 126; dancing not fighting 117–118; Hang 119–121; installation best practice 118–119; troubleshooting 127–129

# M

magic light bulbs 155–157, 160
Master Electrician 5–8, 14, 16, 24, 65
Materials Price Out 63
"Merge Show File" dialogue *58*
Microsoft Access inventory database *168*
multi-cable termination locations *84*

# N

network protocols 90, 91
notes calls 142–146

# O

one-to-one dimmer-circuit relationship 70
opto-splitter 92, *93*

## P

paperwork 110–112
Paperwork Prep 53, 67, 71
patch cables 71, 72
patch sheet 81, *82*, 85
peer to peer collaboration 24–25
Peukert's formula 162
phase balance 79, 80, 86
plastic birthday cake *11*
plot clean-up process 56–60
Plot Review 53–56, 65
Prep Process 53, 61
pre-production process 43–51; design meetings 47–49; onboarding 43–51; plot submission 49–51
price out 61–66
producers *vs.* venue 6
Production Electrician 5–6
production paperwork 177–181
production planning 28
Props Department 12
Props Director 12
prototype 160–161; LED Halo effect 160, *161*

## R

regional theatre, structure *10*
remote controlled birthday cake *158*
resident theatre companies 8
resident theatres 8, 17
review notes 54
rigging planning 110–112
role, defining 30–31
rules of thumb 175
Run Crew 16, 17

## S

season budgeting 171–173
season planning 165–173
Shop Prep 113–115
Show Hat 32, *32*
solution approval 160–161
solutions, budgeting 158–160
solutions, planning 158–160
SOOW 75
Spiider Number 90–91, 95, 96
Stage Management 88, 142, 143, 146
strike 148–149

## T

Team Hat *32*, 33
technical team 47–51
technicians 16, 33, 37, 38, 103, 105, 115, 119, 120, 131–133, 135–137, 139, 146–148, 160
Template Count Sheet 100, 106, 107
template designer onboarding e-mail *44*
temporary systems 68–71, 73
Three Hat Philosophy 32–33
time planning 55, 151
tracking notes 142
troubleshooting 84, 126–129
Truman, Harry S. 27

## V

Vectorworks 43, 56, 59–60, 80, 94, 96, 97, 101, 103, 159